ALSO BY PRISCILLA LONG

Minding the Muse: A Handbook for Painters, Composers, Writers, and Other Creators

Crossing Over: Poems

The Writer's Portable Mentor: A Guide to Art, Craft, and the Writing Life

Where the Sun Never Shines: A History of America's Bloody Coal Industry

(editor) *The New Left: A Collection of Essays*

FIRE AND STONE

John Griswold, *series editor*

fire and stone

Where Do We Come From?
What Are We?
Where Are We Going?

PRISCILLA LONG

The University of Georgia Press
Athens

Published by the University of Georgia Press
Athens, Georgia 30602
www.ugapress.org
© 2016 by Priscilla Long
All rights reserved
Designed by Erin Kirk New
Set in 10 on 14 Minion
Printed and bound by Thomson-Shore

Most University of Georgia Press titles are
available from popular e-book vendors.

Printed in the United States of America

20   19   18   17   16   P   5   4   3   2   1

COVER PAINTING: *Hope* by Jacqueline Barnett, oil, 34 × 36, 2014.

Library of Congress Cataloging-in-Publication Data

Names: Long, Priscilla, author.
Title: Fire and stone : where do we come from? what are we? where are we
going? / Priscilla Long.
Description: Athens : University of Georgia Press, [2016] | Series: Crux :
the Georgia series in literary nonfiction ser. | Includes bibliographical
references.
Identifiers: LCCN 2016016097 | ISBN 9780820350448 (pbk. : alk. paper)
Subjects: LCSH: Long, Priscilla. | Authors, American—21st
century—Biography. | Humanity.
Classification: LCC PS3612.065 Z46 2016 | DDC 818/.609 [B] —dc23
LC record available at https://lccn.loc.gov/2016016097

To my nieces and nephews, with love and admiration.

Dan Long and Michele Cooper-Long

Allison Korn and Marco Yunga Tacuri

Joanna Long and Mike Becker

Eric Messerschmidt

I return where fire has been,

to the charred edge of the sea.

ROETHKE

CONTENTS

FIRE AND STONE

In the Beginning

Where do we come from? What are we? Where are we going? These questions stand as the title of an 1897 painting by Postimpressionist painter Paul Gauguin, and they are the questions that drive this book. For what we are and where we are going has everything to do with where we come from and *who* we come from. "Suddenly all my ancestors are behind me," writes poet Linda Hogan. "Be still, they say. Watch and listen. You are the result of the love of thousands."

The past exists—in cryptic form, in code—in the present. It's here in the rocks, in fossils and landforms, in lakes and mountains and plains. It's here in the structures of bridges and buildings, in current technology, which is based on past technology. The past arrives in light shining from stars that began traveling toward us millions or even billions of years ago. We carry the past in our body, in our language and culture, in our daily habits, even in our loves. We carry the traumas of the past, and if we choose to forget them, our bodies remember them and hold them. Our very chromosomes—all twenty-three pairs—came to us from the deep past, from a long reproductive journey that began with the beginning of life on earth. And too, we keep our childhood into old age, in shrunken, distorted, partly obliterated form. Perhaps we keep it so we can return to its happiness or even to its miseries. "So, like a forgotten fire," writes the French philosopher Gaston Bachelard, "a childhood can always flare up again within us."

Memoir is personal and unique and individual. It's who we are and how we got to be that way, the stories we carry, the roads we've traveled. But our unique selves also carry our inheritance. Begin with the body itself, our animal form, our brain and thus our dreams, our mother tongue. Begin with fire. Our ancestors discovered the use of fire and passed it down to us. We are their inheritors.

We *Homo sapiens* go back ten thousand generations, more or less, but the farthest back I can specify on my mother's Pennsylvania Dutch side is to a horse thief named Christoph Tanger, a German innkeeper who on March 13, 1749, became the focus of a procession through a town named Gemersheim on the edge of the Rhine River. The procession led to the gallows, and at its conclusion Christoph Tanger was hanged. His wife and child came to America, and I am their inheritor. On my father's side there's the Scottish craftsman Andrew Sproul, one of my great-grandfathers, who repaired looms in Scottish weaving factories at the dawn of the Industrial Revolution. There are also the Winslows, English farmers who migrated to New England. I am here because of them.

Fire and Stone bears the mark of one who grew up on a farm on the Eastern Shore of Maryland, raised on old books and cow meat and Robert Frost poems. It bears the mark of one who came of age during the 1960s, the Vietnam War era, with its turmoil and protest, with its battles and burning monks and body bags, so many body bags. It bears the mark of one who spent twelve years in the printing trade in Boston, and in the old-time music scene. It bears the mark of one who ended up in the Pacific Northwest, in Seattle, that city of readers and writers. My works and days have always included reading and writing. My idea of the good life includes a good book and a fire in the fireplace, or it includes a good coffeehouse, a good cup of espresso, and a notebook— yes, that old-fashioned product of the forest industry—to write in.

Who we are includes the landscape our brains have mapped, the ancestors who gave us our lives, the cultural currents that we carry and that burn through us to the future. Each one of us is unique, but more than that we have value and purpose beyond personal strivings, achievements, failures, and fortunes. We carry the flame of life itself, and we carry the flame of art and culture and language. We carry our own part in violence and in love. We carry this torch—whether in awareness or all unaware—for whatever time we are given to be here on earth. And we pass it on.

Fire and Stone is an excavation, an exploration, a tentative tapping of the blind stick toward answers to Gauguin's questions. Where do we come from? What are we? Where are we going? They remain core questions. To follow them will take us not to some final destination but to roads that wind a long way around and lead back to ourselves.

I Inheritance

Look to the rock from which you were hewn,

and to the quarry from which you were digged.

ISAIAH 51:1

Interview with a Neandertal

Each twist of fate may have its interpretation, but it also has its beauty.
JAMES HILLMAN

They were human beings. They were our cousins, even if one thousand times removed. And they were not one thousand times removed. As it turns out, all *Homo sapiens*, except those whose ancestors never left Africa, have a bit of the *Homo neanderthalensis* genome. In 2012 I had my genotype read by the firm 23andMe, and they tell me that my genome is 2.9 percent Neandertal. After years of following the Neandertal story, do I need to tell you that I am thrilled?!

But who were they, really? What were their lives? What if we could return to their time? Here are my questions to them.

1. You Neandertals went extinct perhaps forty thousand years ago. Still, your time on Earth lasted several to many millennia longer than our time has lasted so far. If you could speak to us now, could you teach us anything about survival?

2. We evolved in Africa perhaps 200,000 years ago. Between 195,000 and 123,000 years ago, climate change in the direction of freezing cold wiped out most of the founding population of *Homo sapiens*, according to some thinkers. This calamitous cooling event decimated a population of 12,800 individuals down to a remnant of 600. In this view, we today—all of us—descend from those 600 survivors. Now, as for you Neandertals, we are constantly inquiring as to how (not whether) we

were cognitively superior to you, given that we survived and you didn't. But is that really different from asking how our 600 surviving ancestors were cognitively superior to their kin who perished so long ago? What is it about that question?

3. You *Homo neanderthalensis* and we *Homo sapiens* had a common ancestor, a humbler hominid who spread a long way out of Africa a long time ago. This predecessor hominid lived maybe 440,000 years ago. You Neandertals evolved out of those folks in Europe at some poorly understood time, say about 300,000 years ago. Our people evolved out of the African version of those same folks. Gradually you spread south, and some of you ended up in the Near East. Gradually we spread north and some of us also ended up in the Near East. And there we met, perhaps 60,000 years ago. We occupied the same region for about 20,000 years before you vanished. The longstanding question: Did any of yours get it on with any of ours? This was answered in May 2010 when researchers announced: Yes! Some of yours did get it on with some of ours. This *Homo Neanderthalensis/Homo sapiens* hoochie-coochie occurred after some of us arrived in the Middle East and before we moved off in different directions. We all, except for pure Africans whose ancestors remained in Africa, carry a few of your genes. So, what did you think when we first met? Did you think we looked funny? Did you think we had strange ways?

4. Was it love or was it rape? Or was it some of each? Did we have any Romeo and Juliet–type situations?

5. Did we find each other attractive? Repulsive? Some of each? You had stockier bodies, thicker bones, bigger muscles, a bigger brain. You had a massive eyebrow ridge, that supraorbital ridge. You had positively huge noses with wide nostrils. Did you laugh about us and call us the nose-less know-it-alls? Did we laugh about you and call you the chinless bone-chewers of the North?

6. Once upon a time, a Neandertal man (Shanidar 3) was killed with a spear. He was forty to fifty years old. The time: between fifty and seventy

thousand years ago. His spear-impaled remains were discovered in Iraq in 1959. We know now that spears predate both *Homo neanderthalensis* and *Homo sapiens*, due to three-hundred-thousand-year-old spears found in Schöningen, Germany. So, both your people and our people used spears. Was this death caused by an interspecies conflict, or did another Neandertal person stab him to death? What happened here? Who threw the spear? Why?

7. Many of you had upper-body injuries akin to those of modern-day rodeo performers. Did you fight mammoths at close quarters? And you female Neandertals had powerful wrists and powerful hands and you sustained upper-body injuries, just as your men did. Did you hunt right along with them? If so, who did the cooking? And since we think your children had lengthy childhoods, even if with moderately different developmental patterns, who cared for the children?

8. This is a question about love and death and sorrow and dinner. Upon occasion, you dined upon your own kind. And, upon occasion, we had Neandertal for dinner. (Did you disappear, as some suggest, because we ate you up?). We, of course, have also dined upon our own kind. Did you, also, bury your dead in grief and sorrow? Could it be that in your 250,000 years of existence among a widely spread-out population, there were cultural differences among you? Could it be, in fact, that you were not one people but comprised various bands and groups related closely or not so closely that evolved from different pockets of more archaic humans?

9. Fact: Within caves, *Homo sapiens* artifacts and remains are consistently found above Neandertal artifacts and remains. Interpretation No. 1: We wiped you out. Interpretation No. 2: We couldn't enter your living quarters until you had already quit them. Which is it?

10. We have found, in Iraq, some old bones, the bones of an ancient Neandertal we call Shanidar 1, who was decrepit, who was arthritic, who lived with a fractured eye socket, a withered arm, a crippled leg, a broken foot. But before his death about forty thousand years ago, his

injuries had healed. No way could he have lived in this condition without protection, without care. Did you, then, cherish your elders?

11. None of your buried remains have lower body injuries. Were you nomadic? If an injury immobilized you, what happened to you? Were you abandoned outdoors, where bones don't keep? Where hungry hyenas hunt for hurt humans?

12. You had a hyoid bone—required, along with the tongue, for human speech. You had the FOXP2 gene, the language gene, same as we do. A mutation of this gene causes a *Homo sapiens* individual to face difficulty articulating words with his mouth and tongue and difficulty in language comprehension. Your brain had a Broca's area and a Wernicke's area, essential for speech and for language comprehension. But your larynx was differently configured and you had a larger tongue. Did you communicate in language? If so, what did you speak about?

13. If you indeed communicated in language, which seems likely, in how many tongues did your people speak, considering that you lived for 250,000 years across a territory ranging from Siberia to southern Spain to the Near East? Were any of you bilingual? Did you have creation stories? When we first met, how did we communicate?

14. Until recently we superior beings thought that you did not have the mental goods for abstract thought, that you did not make symbols. We thought that, to the extent you made jewelry (an index of symbolic thought), you got it from us. Until a Neandertal shell necklace was found—a necklace fabricated long before we came along. Along with pigments used for body paint, the earliest ever found. If these finds hold up as trustworthy, if you did have symbolic thought, what were your thoughts?

15. We know that you were fair-skinned (your genome includes the gene for redheads), whereas we, in Africa and recently out of Africa, were darker-skinned. We were thinner and taller, narrower-hipped. We

had smaller brains, although we were equal in intelligence. Did we look down on you because of your pale, sickly looking skin?

16. You hunted, you scavenged, you survived the shocks of ice ages, you survived extreme climate change, you fought woolly mammoths, you fought woolly rhinoceroses, you competed with hyenas for caves, you got hurt, you broke bones, you butchered, you built fires, you cooked, you built bone huts, you sewed mammoth furs with bone awls, you used your teeth as a third hand, you crafted stone tools, you cared for your old, you walked miles per day, you made jewelry, you wore jewelry, you painted your bodies, you spread throughout Europe, you spread into the Mediterranean, you had babies, you killed, you buried your dead, you grieved. But did you tell stories around the fire? Did you relate to your children the tale of all that had happened before?

17. Did you sing? Did you dance? Did you rock your toddler to sleep with a lullaby? What games did your children play? Did they play Ride the Mammoth or Hyenas and Humans? Did they play tag?

18. At the height of your time on Earth there lived perhaps fifteen thousand Neandertal individuals. Close to the number of people who today reside in Tillman's Corner, Alabama. Or Brookings, South Dakota. Or Easley, South Carolina. And your clans lived widely spread out throughout Eurasia. You, in Europe, had to deal with severe cold and worse, with sudden climate shifts. We, in Africa, had it warmer and maybe not easy, but certainly easier. Clive Finlayson argues that climate curtailed your numbers. Friedemann Schrenk and Stephanie Muller note how common extinction is, more likely than not. They think you died out due to a simple failure of reproduction. Fifteen individuals produce seven individuals produce three individuals. Another theory holds that we began interbreeding, and your genome simply got swamped out. What happened? Did you foresee the end of your own kind?

19. I used to call persons I considered backwards, ignorant, small-minded, bigoted, and stupid "Neanderthals." Will you, brainy being of

another time, dreamer of the ice ages, fair-skinned human being of Pleistocene Europe, accept my apology?

20. What did you long for? Did you fear death? Did you love and were you loved in return? Did you name your children? Did you propitiate spirits, imagine your ancestors, keen your dead? If you could speak to us now, what would you say?

2

My Brain on My Mind

The dream of the dead
acted out in me.
MURIEL RUKEYSER

Walter Long was a writer, and he was my grandfather. He was cour-
teous, charming, chivalrous, handsome, well spoken, well shaven, well
dressed, and completely senile. His mental decline began when I was
a girl. In the end, he didn't know me, and he didn't know his own son,
my father. He was born in 1884. He wrote for four or five decades until,
starting sometime in the 1950s, dementia destroyed his writing process.
We have a photo of Granddad writing with a dip pen at a slant-top
writing table. He was a tall, thin man with a high forehead and a classic,
almost Grecian, nose. He was a metropolitan reporter for Philadelphia's
leading newspaper, the *Philadelphia Bulletin*, before the era of regular bylines. What remains of his five decades of reportage? Nothing.
His words have been obliterated, eradicated, annihilated. And what do
we know about his brain? About his neurons, or ex-neurons? Almost
nothing. Before me, my grandfather was the writer in the family. In the
end, what did he know? What did he remember? Did he even remember the alphabet? I dedicate this chapter, shaped by the alphabet, to
him. To his memory.

A.

Alphabets are an awe-inspiring invention of the *Homo sapiens* brain.
Consider these sound symbols lining up before your eyes. Our

twenty-six letters can create in English one to two million words. (The range has to do with what you consider a word. Are "brain" and "brainy" the same word?)

Where in our brain do we keep our ABCs? How does our brain provide us with the use of alphabetic characters without thought? I am handwriting this sentence in my writer's notebook. The letters flow out of my pen as if they were a fluid flowing from my fingertips rather like sweat. Nothing for which I really have to use my brain.

B.

My brain boggles my mind. Its mystery. Its moody monologue.

I walk down Bagley Avenue this fine April day. The Seattle sky is blue. The Brain, wrote Emily Dickinson, is wider than the Sky, since it contains both Sky and You. My own brain contains this blue sky plus six cherry trees in full bloom. Plus the memory of my granddad's face. Plus bungalow yards and rock gardens bright with tulips, violets, camellias, and azaleas. The passing scene enters my eyes in the form of light waves. Neurons in my retina convert these light waves into electrical impulses that travel farther back into my brain.

Our brain contains 100 billion neurons (nerve cells). Our gray matter. Each neuron has an axon—a little arm—that transmits information in the form of electrical impulses to the dendrites—receivers—of nearby neurons. Dendrites branch twig-like from each neuron. Between axon and dendrite, the synapse is the point of connection. Axons commune with dendrites across the synaptic gap.

When neurons "fire," they emit a rat-a-tat-tat of electrical pulses that travel down the axon and arrive at its terminal endings, which secrete from tiny pockets a neurotransmitter (dopamine, say, or serotonin). The neurotransmitter ferries its message across the synaptic abyss and binds to the post-synaptic dendrite, whereupon the synapse converts it back into an electrical pulse.

What blows my mind is this: a single neuron can make between a thousand and ten thousand connections. At this moment our neurons are making, it could be, a million billion connections.

What this electrical/chemical transaction gives us is culture: nail polish, Poland, comic books. Otis Redding belting out "Try a Little Tenderness" at the 1967 Monterey International Pop Festival, along with its memory, its YouTube reenactment, its recordings and coverings and remixings, its moment in history.

The geography of the brain ought to be taught in school like the countries of the world. The deeply folded cortex forms the outer layer. There are the twin hemispheres, right brain and left brain. (We may be of two minds.) There are the four lobes: frontal in front, occipital (visual cortex) in back, parietal on top, and temporal behind the ears. There is the cerebellum ("little brain"), essential for fine motor control and who knows what else. Within the temporal lobe lie the engines of memory and emotion: the hippocampus (memory) and the amygdala (fear). There's the thalamus, central relay station. There's the brain stem, whose structures keep us awake (required for consciousness) or put us to sleep (required for regeneration of neurotransmitters).

The brain has glial cells, once considered mere glue. Glial cells surround and support neurons, carry nutrients to neurons, and eat dead neurons. They regulate transmission of neurotransmitters such as glutamate (excitatory) and GABA (inhibitory). Other glial cells produce myelin, a fatty white matter that surrounds and protects long axons. When stimulated, glial cells make, not electricity as neurons do, but waves of calcium ions.

So there you have the brain: a three-pound bagful of neurons, electrical pulses, chemical messengers, glial cells. There, too, you have the biological basis of the mind. "Anything can happen," says C. D. Wright, "in the strange cities of the mind." And whatever does happen—any thought, mood, song, perception, delusion—is provided to us by this throbbing sack of cells and cerebral substances.

But what, then, is consciousness?

C.

Consciousness, according to neuroscientists Francis Crick and Christof Koch, is "attention times working memory." "Working memory" being the type of memory that holds online whatever you are attending to right now.

And then there's that particular state of consciousness we call self-consciousness—the sense of self, the sense of "I" as distinct from the object of perception. If I am conscious of something, I "know" it. I am "aware" of it. As Antonio Damasio puts it in *The Feeling of What Happens*, "Consciousness goes beyond being awake and attentive: it requires an inner sense of the self in the act of knowing." (It also requires neurotransmitters, glutamate, say, or acetylcholine . . .)

In dreamless sleep, we are not conscious. Under anesthesia, we are not conscious. Walking down the street in a daze, we are barely conscious. Consciousness may involve what neuroscientist Jean-Pierre Changeux postulates is a "global workspace"—a metaphorical space of thought, feeling, and attention. He thinks it's created by the firing of batches of neurons originating in the brain stem, whose extra-long axons fan up and down the brain and back and forth through both hemispheres, connecting reciprocally with neurons in the thalamus and in the cerebral cortex. These neurons are focusing attention, receiving sensory news and assessing it, repressing the irrelevant, reactivating long-term memory circuits, and, by comparing the new with the known, registering a felt sense of "satisfaction" or "truth," which is brought home by a surge of the reward system (mainly dopamine).

Crick (who died in 2004) and Koch proposed, rather, that the part of our gray matter necessary for consciousness is the claustrum, a structure flat as a sheet located deep in the brain on both sides. Looked at face on, it is shaped a bit like the United States. This claustrum maintains busy back-and-forth connections to most other parts of the brain (necessary for any conductor role). It also has a type of neuron internal to itself, able to rise up with others of its kind and fire synchronously. This may be the claustrum's way of creating coherence out of the informational cacophony passing through. For consciousness feels coherent. Never mind that your brain at this moment is processing a zillion different data bits.

Gerald Edelman's (global) theory of consciousness sees it resulting from neuronal activity all over the brain. Edelman (along with Changeux and others) applies the theory of evolution to populations of neurons. Beginning early in an individual's development, neurons firing and connecting with other neurons form shifting populations as they interact with inputs from the environment. The brain's reward system mediates which populations survive as the fittest. Edelman's theory speaks to the fact that no two brains are exactly alike; even identical twins do not have identical brains.

How, in Edelman's scheme, does consciousness achieve its coherence? By the recirculation of parallel signals. If you are a neuron, you receive a signal, say from a light wave, then relay it to the next neuron via an electrical pulse. Imagine a Fourth of July fireworks display, a starburst in the night sky. Different groups of neurons register the light, the shape, the boom. After receiving their respective signals, populations of neurons pass them back and forth to other populations of neurons. What emerges is one glorious starburst.

Or here's another approach to consciousness. Panpsychism. In its looser form, panpsychism holds that every particle—every quark or rock—possesses protoconsciousness. But why would such a thing be so? And how would the protoconscious particles of the universe aggregate to become conscious? Christof Koch proposes a narrower version, in which consciousness is a fundamental property not of particles but of integrated systems (the way a charge is a fundamental property of an electron). Koch works with the Integrated Information Theory worked out by neuroscientist Giulio Tononi. In a brain, populations of neurons firing and connecting with other neurons compose an integrated information system. So do the brains of other animals: in Koch's view, dogs and even worms have some form of consciousness. Not so, computers. A computer may contain billions of bits of information, but the bits—say two photos on a desktop—are disconnected from one another. Well then, is the Internet conscious? Does the Internet, with its zillions of interconnections, feel like something to itself? We do not know.

I myself do not have a theory of consciousness. Still, I am a conscious (occasionally) being. My sense of myself, my sense of an "I" has some sort of neuronal correlate. I am conscious (aware) of the fact that I am teaching a writing seminar (observed object with neuronal correlate) on the literary form known as the abecedarian (observed object with neuronal correlate). I am conscious (aware) that I will be submitting my own abecedarian—this one—to the brainy writers in the class. Because I can imagine the future, because I can plan ahead (thanks in part to my frontal lobes), I feel apprehensive. How crazy! To imagine I could comprehend the *Homo sapiens* brain, the most complex object in the known universe, within the twenty-six compartments of an abecedarian.

I will try. I will color the cones and rods and convoluted lobes printed in black outline in my anatomy coloring book. I will teach my neurons to know themselves. As I write this, I picture our class seated around our big table. I can picture the face of each writer at the table. To each face I can attach a name. This is proof that, as of today, I have dodged dementia.

D.

Dementia dooms a life. It doomed my grandfather's life. Even today, when Alzheimer's disease—just one type of dementia—afflicts one-third of all Americans over the age of eighty-five, according to the Alzheimer's Association, we know far too little about it. It's not clear what kind of dementia Walter Long had. He may not have had Alzheimer's. He may have had Lewy body dementia. He may have had small strokes. Whatever it was, it doomed his brain, it doomed his body, it doomed his body of work, including a novel, never published, that must have existed as a typescript. Upon his death following years of senility, this novel was discarded. For me, the disappearance of my grandfather's writing is a distressing enigma. Not an easy problem.

E.

Easy Problem. Philosopher's lingo for the problem in neuroscience of comprehending the neuronal correlates of consciousness. When you see red, what exactly are your neurons doing? When you remember your grandfather's face, what are your neurons doing? It may be difficult to parse the answer but in principle we can do it. It's easy. The

Hard Problem is the mystery of subjective experience. When long light waves stimulate our neural pathways, why do we experience the color red? Why do we experience anything?

And here's another problem, also hard. What survival benefit caused our brains to develop, through eons of evolution, an ability to experience a "sense of self," a self able to see itself as special or heroic or smart or not so smart—as, on occasion, a complete failure?

F.

Failure to learn new things kills neurons. People who vegetate before the TV are killing their neurons. People who never do anything new or meet anyone new are killing their neurons. People who never read or learn a new game or build a model airplane or cook up a new recipe or learn a new language are killing their neurons. Mind you, many middle-aged professionals are killing their neurons. They're doing what they are good at, what they already know, what they learned to do years ago. They're pursuing careers, raising children, cooking dinner, returning phone calls, reading the newspaper. They are busy and accomplished, but they are not learning anything new. If you are not learning anything new, you are killing your neurons. To keep your neurons, learn something new every day. Begin now. Doing so requires no particular genius.

G.

Genius is nothing you can be born with. No one is born with it. Not Mozart, not Picasso, not Tolstoy. In any field, world-class achievement demands at least ten thousand hours of practice. According to Daniel J. Levitin in *This Is Your Brain on Music*, dozens of cognition studies have produced the same result: geniuses practice more. Neural pathways require repeated stimulation to attain a "genius" level of mastery. The neurons must be stimulated and restimulated, over and again. Essential to this learning process, to this process of achieving supreme mastery, is the hippocampus.

H.

The hippocampus is at the core of what is known as declarative memory—memory of facts and events that can be recalled later for conscious reflection. Memories of what you did this morning, of which candidate you voted for, of whether you were supposed to bring home milk or eggs, all depend on the hippocampus. Alzheimer's destroys the hippocampus. The sea horse–shaped structure, one on each side, is located above the eye, about an inch behind the forehead.

We remember what is emotional. Fear, essential for survival, is provided to us by our almond-shaped amygdalae. Fearful events fire up the amygdala and the amygdala sends its projections all over the brain, but especially to the hippocampus. The amygdala can smell a rat. It receives sensations directly from the nose and sets off alarms with no intervening cognition. We remember what we fear. And we remember what we like, what we want, what we love, what triggers our reward system, dopamine, serotonin. We attend to what is meaningful, what is emotionally resonant, whether positive or negative. We remember what we pay attention to.

Hippocampal activity is not essential for procedural memory—what the body knows. You don't need your hippocampi to ride a bike or get out of bed or even play the piano if you are a pianist. The hippocampus is not essential for semantic memory—facts and words. It's not even essential for working memory—remembering a phone number long enough to punch in the numbers. But it's the brain's transformer of short-term memory into long-term memory. What you lose when you lose your hippocampi is your ability to make new long-term memories.

Such was the fate of the much-studied HM, Henry Gustav Molaison (1926–2008). His tragic case gave us much of what we know about memory. In 1953 a neurosurgeon, attempting to halt the young man's frequent epileptic seizures, removed most of HM's hippocampi, his amygdalae, and some surrounding tissue of the temporal lobe. The seizures stopped. And HM could still speak and make perfect sense (semantic memory). He could remember his old skills and even learn new skills (though he couldn't remember learning them). He retained long-term memories, including vivid childhood scenes. He retained his high IQ. What he lost—in terms of a life, almost everything—was the capacity to turn new short-term memories into long-term memories. He could not remember what happened yesterday. He could not remember what happened this morning. He could not remember the scientists who studied him for forty years; he met them anew at each encounter. After the surgery he could no longer care for himself and lived in a nursing home. "HM's case," wrote neurologist Oliver Sacks in *Musicophilia: Tales of Music and the Brain*, "made it clear that two very different sorts of memory could exist: a conscious memory of events (episodic memory) and an unconscious memory for procedures—and that such procedural memory is unimpaired in amnesia."

The conscious memory of events: how we take it for granted! It enables us to plan, to pursue a goal, to work, to cook, to read. It enables us to enjoy long talks and lazy days and nights out on the town. It enables storytelling, art, imagination.

I.

Imagination depends on the conscious memory of events. How could I imagine a purple cow if I could not remember the cows of my childhood switching their tails against the horseflies? How could I imagine a purple cow if I could not remember purple hearts, purple grape juice, the purple shawl I knit for my Grandmother Henry—my mother's mother? Persons with impaired memories have impaired imaginations. Amnesiacs, writes Benedict Carey, "live in a mental universe at least as strange as fiction: new research suggests that they are marooned in the present, as helpless at imagining future experiences as they are at retrieving old ones." Images made by functional magnetic resonance imaging (fMRI) technology show that remembering and imagining send blood to identical parts of the brain.

What does this say about the goal of living in the present?

But for most of us, the phenomena of the present (just now, Miles Davis playing "Red China Blues" on YouTube) connect in our mind with previous analogous experiences. Recognition involves memory: comparing what is seen with what was seen.

My grandfather had, I think, anterograde amnesia: He couldn't form new memories. He could remember the long ago but not yesterday. He would get dressed in his suit and tie, don his fedora, dapper as ever, and head out the door.

"Walter! Where are you going?" Gran would ask.

"I'm going to work," Granddad would say.

"You're not going to work!" Gran would cry out. "You're retired!"

Granddad lived, I think, in a state of perpetual churning anxiety. He felt it was time to go to work. He felt lost. He wondered out loud who these "nice people" were, sitting in his living room. (That would be us, his family.)

In the process of losing his memory, did Major Walter Long lose his pride in being decorated for "exceptional bravery under shellfire" in 1918 France during the Great War? Did he forget the trauma of war, his killed comrades? Did he forget the pleasure of composing a paragraph? Did he forget love? Did he forget joy?

J.

Joy, happiness, contentment, the feeling of safety, the feeling of being loved, the act of loving, the feeling of respecting another and of being respected, all these feeling states are produced within the brain. The pursuit of happiness might be construed as the pursuit of more dopamine and/or serotonin flooding our synaptic clefts. Add norepinephrine to the mix—energy, the constricting of blood vessels, jumping up and down. Norepinephrine is a hormone when produced by the adrenal gland along with epinephrine (adrenaline). It's a neurotransmitter when produced by neurons in the brain. Certain racers, bikers, fistfighters, bank robbers, pickpockets, and other daring devils may be addicted to the intoxicating rush of norepinephrine-epinephrine.

Normally, these neurotransmitters spread out and do their job, after which they break down within the synaptic clefts or are returned to their home neurons by reuptake molecules. Antidepressants like Prozac or Zoloft (SSRIs—selective serotonin reuptake inhibitors) bind to serotonin reuptake molecules, preventing them from doing their ferrying duty. This leaves serotonin flooding the synaptic gaps, free to continue stimulating the receptor molecules in the dendrites of the receiving neurons.

Cocaine binds to both serotonin and dopamine reuptake molecules, leaving the synaptic gaps awash in both. Whee! But then the crash. Receptor molecules in the dendrites are switches. When stimulated they switch on; when overstimulated they switch off. (With his or her receptors desensitized, the addict needs more and more.) And, because the neurotransmitters never get returned to their neurons, the dopaminergic and serotonergic systems get depleted, drawn down, drained out. Quite soon the system itself becomes deranged. Many addicts, whether using or recovering, have damaged brains. Tragically, lacking crack, they can feel no pleasure.

I'm no addict, but I do get migraines. This means I likely have a low supply of norepinephrine, an excitatory neurotransmitter that counterbalances dopamine. Under migrainous conditions, dopamine flooding my synaptic clefts leads not to a high but to the worst kind of low—killer headaches.

K.

Killer headaches—including nausea, vomiting, light-stabs to the eyes, repulsive odors, excruciating head pain, a sense of total despair—are under study by me when I'm not having one. Migraine is cousin to epilepsy. It may be in part genetic, although Pamela, my monozygotic twin sister, does not get them. (Mother got them, however.) Migraine begins with an electrical storm in the brain stem, seat of the autonomic nervous system, controller of heartbeat and sleep, dilator of pupils, regulator of airways. This brainstorm spreads widely throughout the brain. Firing neurons require oxygen, carried by blood, and during the brainstorm, three hundred times the normal amount of blood rushes to your head. Now, we migraineurs (according to researcher Stephen J. Peroutka) possess an insufficient supply of norepinephrine, not only during the dread headache but also all the time. Firing neurons secrete norepinephrine, which constricts the blood vessels in the head. So far, no pain. But, alas, our meager supply of norepinephrine gets drawn down, and dopamine (along with its rogue co-conspirators adenosine and prostaglandin), which acts oppositely and in balance with norepinephrine, runs amok. Dopamine distends cerebral blood vessels, which activates the trigeminal (cranial) nerves. Excruciating pain. Dopamine also stirs up the neurons in the stomach lining (we have 100 million of these), creating nausea leading to violent retching.

Another thing about the bodies of migraineurs: we have too much calcitonin gene-related peptide (CGRP). A peptide is a string of amino acids, usually smaller than a protein (proteins are made of amino acids). How CGRP relates to the horrible headache is unknown, but somehow it comes with it, and new (in 2015) drugs to block it are in trial. For about 15 percent of migraineurs in the trials, their headache days are over. Thank you. But what about the remaining 85 percent? And what about me?

Triggers of the killer headache: too much sleep, too little sleep, dark microbrews (the more delicious, the more deadly), too much company throughout a long day, most red wines, MSG, air travel, dark chocolate combined with red wine (requires immediate hospitalization), too much caffeine, too little caffeine. Some women get migraines in sync with their menstrual cycle. Pickles will do it. Sulfites, sulfates, sunlight. Too much exertion. Too little exertion.

Mostly I adore Oliver Sacks's disquisitions on the brain, but I ingested his tome *Migraine* with flutters of anxiety. Might *Migraine* trigger a migraine?

Sacks inquires: What is the usefulness of the migraine to the migraineur?

Well!

There's the alleged migraine personality. Migraineur Joan Didion speaks (in "In Bed") of the compulsive worker, the perfectionist writer. This is the type who slaves over sentences that nonetheless ooze mediocrity like a bad odor. That would be me.

A migraine forces you to stop. Your day ends—bam! A migraine performs approximately the same service as being run over by a train.

Sacks thinks the profound despair brought on by a migraine is part of the migraine, the result of neurons firing out of control.

But what sets off the brainstorm? Why do triggers differ from one person to the next? Why do migraines occur on only one side of the head? And why does my personal miracle drug, Maxalt (rizatriptan benzoate)—which binds to serotonin receptors, which then release serotonin, which constricts blood vessels—cost seventy dollars per headache?

And why me, Lord?

Is my brain sending my body some sort of sick, twisted message, some sort of poison-pen letter?

L.

Letters—our ABCs—are meaningless squiggles until we learn our alphabet. Here's a letter I remember. I'm four or five years old. I'm sitting on the davenport in the living room. I'm holding this letter in my hands. Pale blue letter paper. Blue ink. Gran, my Scottish grandmother, has written this letter to Mummy. I turn it over. I turn it around. I turn it every which way. I put it close to my face. I hold it far from my face. I turn it upside down. I'm filled with longing. I long to know its secrets. I long to read this letter. But I cannot read. Mummy comes into the living room and takes the letter from me. Foiled! And with the letter she takes the letter's letters. I feel completely exasperated!

What part of the brain does this desperate desire to read come from? And where does the brain keep it—the long-since-satisfied longing to learn to read retained as a memory?

M.

Memory is nothing like a scrapbook, a photo album, an attic, or a file cabinet. Think of a broom. Remember broom. Different bits of the brain's broom are stored in different parts of the brain. The hickory broomstick. The weight of the broom in the hand. The straw head. The color of straw. The sound of sweeping. The purpose of sweeping. The sound of the word "broom." The shape of the word "broom." The fact that a broom is a cleaning tool and not a glass of wine or a plate of spaghetti. (Thoughts of sweeping, for those who sweep, activate a pre-motor area, ready to lift the hand.) Memory brings all these disparate bits together, makes them cohere. The puzzle of how the brain achieves coherent perceptions out of its widespread data bits is known as the binding problem.

Memory is a mental event, this we know. Mental events work by the transmission of neural impulses at different rates. Memory is stored not in one place but all over the place, as data bits. Memory, says Antonio Damasio, likely involves "retro-activation"—the refiring of neurons activated during an original perception or experience. An association, either external or from within, may stir up a memory.

Types of memory: procedural (how to sweep the floor); semantic (facts, words, the word "broom" and to what it refers); working (being used at this moment to consider the concept of a broom); episodic (personal memories, the time you swept up your diamond ring with the dirt); declarative (remembering facts and events that become available for later conscious reflection).

Lost to everyone's declarative memory is the name of Walter Long's first wife, a girl he married in 1914 when he was thirty-one. This girl died of tuberculosis a year or two after she married the young man who would become my grandfather. After her death, Walter went off to fight in the Great War. He was proud of his service (my father said). He received the Croix de Guerre. Toward the end of the war, he got the mumps, requiring nursing. In 1919 he married the sister of his nurse, a young Scottish war widow with a small child. This young mother, Annie McIlwrick Sproul Humphrey, became my Gran.

But Granddad's first wife—who was she? When I asked around a few years ago, no one in the family could remember her name, if they ever knew her name. She had gone from this world, gone from memory, gone from history. This girl, whoever she was, went from being somebody—with her looks, her likes and dislikes, with her passion for Walter Long, with a favorite pair of boots perhaps, or a love of pickles—to being nobody. Her dreams died along with her name along with her neurons.

N.

Neurons commune with other neurons. But keep this in mind: a straightforward algorithmic connection from A to B to C is not enough for the brains of human beings or other beings to learn from experience. Rather, neurons act in assemblies that have subsets called cliques. Shifting perceptions are made by shifting transitory assemblies of neurons. In one type of assembly, various neurons receive input at the same time and send their output to the same place. In another type, neurons in different locations fire simultaneously. Assemblies often stack up in columns, with a single column containing perhaps a hundred thousand neurons.

Cliques compete with other cliques, recruiting neurons and losing them to the competition. Let's say you are trying to remember a name, but the wrong name comes to mind. The rogue clique, the clique pulsing the memory of that wrong name, is in competition with the clique you want-want-want-want. Eventually you dredge up the right name from the mind's murky sea. Attention is the net. Attention may be a function of feedback loops ("reentrant connections"), neurons firing from the frontal cortex back to the sensory relay station, the thalamus, to suppress irrelevant stimuli.

Certain neurons work as feature detectors. Neuroscientist Joe Z. Tsien and his team subjected a mouse to an earthquake while recording the activity of some two hundred hippocampal neurons. (Their ingenious lab inventions enable them to observe very few neurons at a time.) The earthquake caused the rodent's neurons to fire in a particular pattern, with different cliques reacting to different aspects. There was a startle clique, a motion-disturbance clique, a clique that reacted to where this event took place (a black box). The startle clique and the motion-disturbance clique both fired again when a different event (an elevator fall) occurred when the mouse was in the same black box. The cliques of firing neurons were organized in a hierarchy from abstract to specific. (Startle is abstract: any number of different events could fire this clique. *Where* is more specific: a red box will not excite black-box neurons.) Memory occurs when, after the event, the same assembly of neurons refires, although less strongly.

The mystery is this: Where does the sense of mystery come from? What about tranquility or annoyance or curiosity or philosophy? Which neurons project ambition or fascination or frustration? Where does the sense of awe come from? What about the sense of the sacred, the sense of God or of *deus in res*? Are these states of being a matter of brain chemistry? Are they nothing more than electrical charges pulsing, thrumming, oscillating?

O.

Oscillating is what the living brain does. It emits brain waves. Neurons emit electrical charges in a rhythmic pattern; they fire even with no stimulation from the outside world. The brain puts out its own energy. I think of this-this-this-this as a kind of humming. Hooked up to the electroencephalograph, electrodes affixed to the scalp, the sleeping person's brain discharges mainly high-amplitude, low-frequency oscillations in the delta band (0.5 to 3 cycles per second). The awake but calm, resting, or meditating person's neurons tend to discharge alpha waves (8 to 12 cycles per second). Beta waves (15 to 25 cycles per second) begin when initiating purposeful activity. The gamma band (30 or more cycles per second) is linked to cognitive activity. But, like a great many statements about how the brain works, this one is oversimplified. In actuality, different brain areas are thrumming at different rates simultaneously. In actuality, the brains of some meditating persons are not in alpha. In actuality, the brains of persons in a TV-watching stupor are in alpha. In actuality, the electroencephalograph gets a lot of interference: with its electrodes stuck not on the brain but on the scalp, it may be a dull instrument. A thick, delicious book, James H. Austin's *Zen and the Brain*, states that more important than the alpha state is synchronicity. Different parts of the brain begin oscillating in unison like the Rockettes at Radio City Music Hall. Bliss may result. But how little we know: our brain has barely begun to comprehend itself. And how wrong it can be. Until the late 1990s the dogma prevailed that neurons do not regenerate, that brain injuries are more or less permanent, that a devastating stroke represents irreparable loss. Then, in the late 1990s, a new insight hit neuroscience like a tsunami: the brain's plasticity.

P.

Plasticity brings hope to the stroke victim, the brain-injured, the autistic, the amputee in phantom pain, the palsied, the deranged, the old. The brain is plastic, not fixed. Brain structures do not have rigid job descriptions. Brain maps—those synaptically interconnected networks of neurons whose pulses produce a function or a memory—have shifting borders. Also, stem cells exist within the brain, particularly within the hippocampus. Brain stem cells can generate new brain cells, perhaps maintaining a balance with dying cells. Plasticity has exploded our notions of how to rehabilitate a stroke victim. Edward Taub, working on macaques, discovered that when one hand is disabled, say by stroke, the brain map for the good hand begins to expand. It is precisely this—the brain's compensatory ability to remap itself—that dooms the paralyzed hand. Taub's strategy is to render the good hand moot by confining it to a sling, and then to force the paralyzed hand to practice—to pick up and drop, pick up and drop, pick up and drop—beginning at the baby stage, putting square pegs into square holes eight hours a day. In this way, new brain maps form in remaining healthy tissue to work the limp hand. Taub's results, according to Norman Doidge in *The Brain That Changes Itself*, have ranged from good to spectacular. Plasticity means that old people can learn, that slow people can raise their IQ, that memory loss can be prevented or reversed.

Learning changes the brain. Gary Wayman and his team discovered that dendrites contain a growth-inhibiting protein. Synaptic activity (learning) moves that protein out of the way. Synaptic activity (learning) also makes the neuron manufacture an RNA molecule (micro RNA 132) that suppresses the manufacture of more of the growth-inhibiting protein, allowing the dendrite to grow. Learning changes the brain by making new pathways and also by growing new dendrites. And cognitive activity, according to psychopharmacologist Stephen Stahl, is the only intervention known to consistently diminish the risk of Mild Cognitive Impairment or Alzheimer's or to slow their terrible progression.

Then again, the propensity to develop late-onset Alzheimer's has a powerful genetic component. On chromosome 19 there's a gene (the

E gene) that codes for a glycoprotein (a protein containing a carbo-hydrate) whose work involves cholesterol transport and metabolism. When it works, it cleans out those waxy amyloid plaques that otherwise clog thoughts and kill neurons. Persons born without a certain allele (alternative form) of this gene (the allele termed ApoE4) are in little danger of developing Alzheimer's. Persons born with one copy of this awful allele are four times as likely to get Alzheimer's as compared to the general population. Persons born with two copies of ApoE4 are eight times more likely to develop Alzheimer's. Very well, but here's the question: What is different about persons who carry two copies of ApoE4 (the worst case) who do not develop Alzheimer's?

And there are other questions.

Q.

Questions. What is it about our brain that makes us human? What is it that makes us different? Is it self-knowledge? Is it, as neuroscientist V. S. Ramachandran puts it, that we have a self that is self-reflexive, a self aware of itself? Is it knowing who or what we are? Is it our ability to explore our past and to imagine our future? Is it our spirituality, our brain's ability to imagine a soul, a higher being? Is it our propensity to make music, to make poetry? And what if we lose all of it, as Walter Long did? What if we lose all that seems intrinsic to our human nature, to our own selves? Who are we then? Who are we if we can't remember?

R.

Remember as you would be remembered. In 2007 my father, Winslow Long, in the process of moving to Seattle, passes on to me a box of old letters and documents. In this battered cardboard carton I discover a booklet titled *The Family Records of Winslows and Their Descendants in America.* We Longs descend from the Winslows. The yellowed, shiny pages of this booklet reveal that my grandfather Walter Long (son of Clara Winslow Long) married Lillian Gorsuch, of Baltimore, on June 10, 1914. I hereby restore to everyone's neurons the name of Granddad's first wife. Did Lillian have tuberculosis at the time of their courtship? Did they know it? Did the coming war in Europe during the 1914 summer of their wedding cause them to feel anxiety? Distress?

S.

Stress shrinks the brain. Not normal stress or necessary stress, but chronic stress—chronic anxiety or clinical depression. That chronic stress destroys dendrites and even entire neurons, that it damages neural pathways, especially in the hippocampus, is the view gaining acceptance as studies go forward.

Stress revs up the adrenal gland to pump glucocorticoids such as cortisol. Cortisol sparks the production of epinephrine (adrenaline), which tenses muscles, narrows blood vessels, and prepares you to kick butt or run for your life. But then the emergency ends and cortisol subsides. All is well. But in chronic stress, the emergency never ends. Cortisol bathes the hippocampus continuously, killing its neurons.

And there's more. The brain produces a protein known as brain-derived neurotrophic factor (BDNF), which protects neurons. Chronic stress may repress the gene that expresses BDNF. After which hippocampal neurons, which thirst for BDNF, which require BDNF, which can't go on without BDNF, shrink or balk or die. Experimental animals subjected to stress, according to *Stahl's Essential Psychopharmacology*, turn off their genes for BDNF and as a result lose synapses as well as whole neurons.

On the other hand, exercise stimulates the growth of BDNF. So insists the molecular biologist John Medina in *Brain Rules*.

So get out and walk. And stop your constant worrying. Stop stressing out over every little thing. Stop imagining the worst. Dementia begins there.

T.

"There is no need for temples; no need for complicated philosophy. Our own brain, our own heart is our temple; the philosophy is kindness." So says the Dalai Lama. But in our world, violence, murder, war, and torture may be as common as kindness. Perhaps we have a deep inner need to kill, a devil in our unconscious.

U.

Unconscious memories, unconscious wishes, unconscious fears, hates, loves. The very notion is strange. Strange to think that we have memories we can't remember, wishes we don't wish for, desires we don't feel. But that we have an unconscious is told by our brain's brilliance at doing things with no help from our conscious mind. We walk, chat, purchase potatoes, sweep, drive, read, talk on the phone, all without "thinking." We just do it. Our brain directs the process, whatever the process is. We have reactions to people and events—a sudden mistrust or a sudden affection—that may be based on implicit, that is, nonconscious, memories of something similar. The admonition "trust your gut" translates as "trust your brain, trust its implicit memories."

Blindsight also argues for the existence of the unconscious. Blindsight proves that we do not necessarily know what the brain knows. A blindsighted person is a brain-injured person. This person's visual cortex has been damaged. He is blind, in his own opinion. Yet ask him to take a guess as to where some particular object is—say a pencil held up—and he will point right to it. The brain sees it. The brain knows where it is. But to the conscious mind, it is unknown. What is broken is the wiring that connects the part of the brain that sees to the part that knows it is seeing. To the person, the world has gone dark. To his brain, the world remains a carnival of shape and color—visual.

V.

Visual arts are unique to our species. By means of culture we have created an external visual cortex—paintings, sculptures, billboards. We have created an external long-term memory—writing. We have created external dreams—films, plays, TV dramas. We tell stories to recall the past and we look through telescopes to see the past. We write in part to stop time, to hold onto the present as it becomes the past as we grow into the future.

I can picture my grandfather's face. I can remember, just barely, a time when he could still be counted among the cognoscenti. He had retired with our Scottish grandmother to a Bucks County, Pennsylvania, farm, the old farmhouse built of whitewashed brick. Granddad used to take us small children out to the barn to show us a sleek black buggy, polished but parked in desuetude. I can see in my mind's eye the barn, the buggy, the big doll I was allowed to play with.

I see Walter Long's life as a tragedy, but maybe he didn't see it that way. He had good work while he could do it and he had love and ambition and at least some of his dreams came true. He reportedly reported on the sinking of the luxury ocean liner ss *Morro Castle* in 1934 and on the Lindbergh kidnapping trial of 1935. I once spent three days searching the *Philadelphia Bulletin* amid the massive coverage of the Lindbergh tragedy for any sign of my grandfather's hand. No luck. But some years later he himself was featured in the paper, in a sidebar, with his picture, here quoted in full (the ellipses appear in the original):

> Walter Long . . . The Zoning Board of Adjustment goes into session . . . hearing pro and con on whether a new apartment site shall be approved . . . News is being made . . . and Walter Long's there . . . accurately recording the builders' arguments, the opponents' vigorous stand . . . For 15 years Walter Long has been one of *The Bulletin*'s experts in municipal affairs. . . . He roams the City Hall annex . . . drops in daily on the Board of Health . . . keeps tab on the Department of Supplies and Purchases . . . and distinguishes himself with his detailed reporting of the City Housing Rent Commission Hearings.

There he is. My grandfather. Not in his own words but in someone's words. Kind reader, if you were to utter the name Walter Long, it would stay longer in this world. It would enter into your Wernicke's area.

W.

Wernicke's area is where the brain comprehends and interprets language. Persons with damage to their Wernicke's area (who have Wernicke's aphasia) can speak, but their words pouring out make no sense. Neither do these persons comprehend a single word spoken to

them. Broca's area produces spoken speech. Persons with Broca's aphasia may be able to speak within their own minds, but when they attempt
to voice their thoughts, they fail to produce normal speech. Wernicke's
area is associated with hearing, whereas Broca's area is associated with
the neurons that activate the muscles of the larynx. Relations between
Wernicke's and Broca's areas are intensely xenial.

X.

Xenial (pronounced ZEE-nial) relations, friendly communicating relations, transpire among many neurons throughout many parts of the
brain. Consider the binding problem, worked on most brilliantly by
psychologist Anne Treisman. As we know, different aspects of the scene
before us are carried into our brain by different neurons. Some neurons signal red; others black or yellow; others the news that what is
before us is vertical or horizontal; others that an object is located in
our upper-right quadrant or our lower-left quadrant. How then do we
reconstruct a coherent picture? How come, when we see a black-and-
white cow with a red ribbon around its neck, the cow doesn't come out
red, the ribbon black and white, since separate neurons have projected
separate features of this beribboned bovine into our brain? The answer
comes from the observation that persons with stroke-injured parietal
lobes may indeed see the cow as red, the ribbon as Holstein. Think
of it this way: it's spatial attention that puts the red on the red ribbon
(both originate from the same point in space). Spatial attention ema-
nates, it seems, from the parietal lobes. Red-perceiving neurons and
ribbon-perceiving neurons are getting together, communing, enjoying
xenial relations rather like people at a cocktail party going yakety-yak.

Y.

Yakety-yak. We are a yackety-yak people. We are quidnuncs, busybod-
ies. Who did what to whom, who went out with whom, who slept with
another's whom, who won the lottery, who won the game, who lost his
shirt. Gossip, it turns out, takes up more than half of all human dis-
course. We concern ourselves with the business of others, and others

concern themselves with our business, and all this sordid business is aired on reality TV, not to mention in cafés and over dinner and upon falling asleep and during morning coffee and later at the bar. Yakety-yak. We social primates evolved within an increasingly elaborate social framework, much dependent on our frontal-lobe-located mirror neurons. When you smile, I want to smile. When you cry, I want to cry. When you laugh, you activate my funny bone. We are inherently at home in social interaction. We can gossip for hours, even if doing so drops our goals met for the day to zero.

Z.

Zero is another awe-inspiring invention of the *Homo sapiens* brain. Zero is intrinsic to our human society, though we seldom give it a thought. All by itself, zero is nothing. So when does nothing become something? Nothing becomes something when you put it next to a 1, as in 10. Now this nothing is holding a place for nothing in the units place. Then if you put two nothings together with a 1 to make a 100, your little nothings are suddenly holding two places: a place for nothing in the units place and a place for nothing in the tens place. The zeros make the 1 mean not 1 but one hundred ones. Think about it. That little nothing, zero, put with only nine other numerals, makes possible any number of numbers. The story of zero is a *Homo sapiens* story, invented by the Sumerians in ancient Babylonia and again by the Mayans in ancient Mexico.

Now, we also have other sorts of zeros. We have ground zero. We have zero population growth. We have the number of extant sentences written by Walter Long. Zero. No paper with Granddad's handwriting on it. No paper typed by him. No article bylined by him (at least none I've found). So here was a writer, my grandfather, who wrote for five decades, who lost his memory, who lost, with his memory, his entire output.

How could this have happened? It's a mystery I ponder even as I hoard every word I write, even as I donate my own scribblings—seventy boxes as of this writing—to a university archives, even as I try to write more and more each day, as if that would overcome the oblivion that for certain lies in the future. Considering everything—wars and famines, families, feasts, births and deaths, the great loves and the great losses, considering the miracle of natural selection that evolved our brains—the loss of my grandfather's writing is a small thing. But for me it's a big thing. I can't get it out of my mind. It leaves me speechless, notwithstanding the two-million-word capacity of our alphabet.

3

Déjà Vu

Past is prologue.

SHAKESPEARE

The past, remembered or not, pulls on a life with the force of gravity. What the mind forgets, our behaviors, our compulsions remember. My father's hands are thick with work, and he works from morning to night. He milks cows, spreads manure, pitches silage from silo to trough. He treats mastitis, washes milk cans. In his spare time, he regrades the long dirt road. He needles the negligent gentleman farmer who is his boss, urging him to give the go-ahead on the plowing or disking or planting. To us, my father says, "If you can't do a job right, don't do it at all."

I am my father's girl. I feed the calves, shovel manure, hose down the gutters in the milking barn. Summer evenings, after the barn work is done, my father cultivates between rows of beans. I mow the grass. I memorize the names of weeds just as my father does—shepherd's purse, purslane, pokeweed, pigweed, dock, spatterdock, buttercup, beggerweed. The fireflies come out and I quit mowing and whisper secrets to my sisters. Night falls. My father walks his dogs down the long dirt lane. The Milky Way swirls great drifts of stars, brilliant in the moonless night.

I grow up into the Vietnam War and spend my twenties needling the government to stop the war. I work from morning to night. I attend meetings and teach-ins and forums and further meetings. I stand on street corners plying passersby with antiwar leaflets. I knock on doors to talk about the war. I wait tables, hawk newspapers, pose for art classes, clean houses. Finally, I become a printer. Operating the press, my hands grow stubby and gnarled, black under the nails, just like my father's hands. By the time I turn forty, I am becoming an old printer with stiff shoulders and thick hands and a permanent frown in my forehead. I crank out job after job, double time, overtime, all the time.

We believe we are making choices, especially in our rebellious youth. We believe we are unique, and we plan to be different, especially different from our parents. If my parents were granite, stony and taciturn, I would be a burbling brook, chattering without cease. If my parents burned with coals of rage, I would be calm and kind, cousin to the color blue. If they went along with the government, I went against the government. In adolescence, I did not get along well with my father. He favored my twin sister. He opposed my wish to be a writer. I forgot that my father's father, senile by the time I was eight or nine, had been a writer for the several decades of his lucidity. I was an unwitting recruit into some sort of murky, undeclared family war. Is it any wonder that I opposed the Vietnam War?

I remember my father's silence. He was silent with his dogs, silent with the cows, silent in a distant field, a lone figure toiling like some sort of bony Hercules. But there were intermissions, lapses into yelling. We are in the milking barn, my father and I. My father is trembling with rage, yelling in the empty barn, about what I no longer remember. The stanchions stand wide open like a child's eyes. My father throws a hammer and it thunks against the cement-block barn wall. My father never hits me, but his rage laps around me like fire.

He teaches me mainly how to work hard. Growing up into the sixties, I shut the door against love. I know nothing about it, it is not my country, I do not speak its language. The continent of touch is unknown to me. I know how to read. I know how to work. I learn how to write. I shun love until it is too late, until it is raging through the Ohio night, raging under dark summer trees whispering in the moonlight. Love sweeps over me like a firestorm, or like a tsunami breaking on the village of the self, smashing it into bits. I speak to my lover, I whisper to him as to my sisters, but my lover turns a deaf ear.

My father began to go deaf in his early thirties. At the time I was eleven or twelve. The three bones in his middle ear—hammer, anvil, and stirrup—began to calcify. Called the ossicles, they are the body's tiniest bones. They amplify sound vibrations by ten times and pass these vibrations to the cochlea, which convert them into electrical signals and send them to the brain. My father inherited his deafness from his Scottish mother. The bones get harder and harder until finally they do not vibrate at all. There is an operation to replace them with plastic ossicles. If the operation succeeds, perfect hearing is restored. But if it fails, the hearer is left altogether without ossicles—stone deaf. Therefore the operation is routinely delayed until little would be lost if it fails. We children have soft voices. Our father shouts at us. He shouts that we are talking behind his back. In the barn he laughs. He says, "Why should I wear my hearing aid in the barn, when there's nobody here but the cows?" We kids stand in a quiet row in our shorts and T-shirts and bare feet.

I talk to my lover in a torrent of words. On the days he is absent I write him long letters, handwritten on one side of lined tablet paper. I consume whole yellow pads writing to him. I post bulky envelopes. After a long time, I realize that these letters are too much for him. They exhaust him and he cannot read them. Once, at his apartment, I saw one of my letters on a dusty end table in the living room. It had not been opened.

That part—the part about the unopened envelope—is not true, but it makes a good story. In the film about my life, the girl enters her lover's apartment one romantic evening and the camera malingers on the living room to reveal evidence of dissolution—dirty underwear, dirty dishes, junk mail heaped and scattered about. On a chipped walnut end table, along with an ashtray clotted with cigarette stubs, she finds a fat and familiar-looking letter. She turns it over—she has polished her long fingernails pink for the shot—and sees that it is her letter, addressed to her lover in her own neat hand, and that it has not been opened.

The reality is that my father is a man of principle, a religious man, highly respected in the farm community in which he is rooted. The town lies a few miles below the Mason-Dixon Line. In the late fifties there are two schools—one black and one white—and two school buses. Some years after the *Brown v. Board of Education* school desegregation case, one little black girl, the token black child, is required to ride the white school bus to the white school. She is in the second grade, in my little sister's class. For the hour it takes the school bus to pick up all the farm children on Quaker Neck and get to school, and for the hour it takes to return all these children to their farms, this little black girl is taunted, jeered, heckled, and tormented. My little sister tells my father. The next day my father visits every farm out on Quaker Neck Road. "It stops," he says at every farm. "And it stops now." It does stop.

Past is prologue. As my father is a man of unyielding principles, so is my future husband a man of unyielding principles. Where I am, I have already been. Where I stand, I have stood before.

The war in Vietnam is escalating. My hero, the love of my life, has received his draft card in the mail. This is some years before we meet. He returns the card to the Cincinnati draft board with a polite letter reviewing his reasons for opposing the war. It is an unjust war. We have not been attacked. The war has never been declared. Besides, there is a legitimate question about what the Vietnamese people do want. We Americans believe in self-determination, don't we? And so on. The Cincinnati draft board responds to his letters by drafting him. He refuses to appear for his physical and a month later refuses to appear for induction. He loses his case, appeals the case. This is approximately when we meet. He loses the appeal and is incarcerated for two and one-half years. I visit him faithfully twice a month, as permitted by the warden.

He gets out of jail, and we get married, and he begins writing a book and I begin writing a book. I am writing a book on the rock that burns, on coal, on the history of coal mining in the United States. I am drawn to the past, fixated on the past. I write in the dark before I leave for work in the morning. I devote Saturdays to technical manuals written circa 1869 on how to sink a mine shaft, how to undercut the coal, how to drive a manway. I crank microfilm reel—*Workingman's Advocate, Shenandoah Herald, The Miner's Magazine.* I record the echoes of miners' voices, Scottish voices, Welsh voices. My father is not a miner, but sometimes these old British immigrants speak in my father's voice. This

coal miner writing a letter to the *Workingman's Advocate* sounds to me
like my father, although I cannot really say why:

> A coal miner, though half his life may be spent in darkness, away down
> in the bowels of the earth, where one ray of God's blessed sunlight never
> shines, is of ten fold more benefit to the world than all of the bankers,
> stock brokers, railroad kings, and other public thieves who are eating the
> very vitals of the nation . . . who are, through their own self-styled aris-
> tocratic, bigoted, selfish, unprincipled actions, hurrying themselves . . . to
> the very abyss of destruction, who by their greedy, avaricious, insatiable
> desire to grasp the accumulations of labor . . . are driving cold steel to the
> heart of every workingman . . . the very creators of all their wealth and
> power. (May 3, 1873)

Every vacation, I travel to Charleston or Omaha or Trinidad,
Colorado. In Washington, D.C., I find an 1875 letter book containing a
mine boss's handwritten daily reports. The newly hired archivist and I
find the thick crinkly book lost among tobacco leaves in a sub-drawer
of a sub-basement of the National Museum of American History. I do
my research with complete thoroughness, as if my time frame were forever. I work like a peasant, without laborsaving devices or laborsaving
wishes. Twenty years later, the same year as my divorce, I publish my
book. I have done the job and I have done it right.

Meanwhile, my father decides to go to college. He has lost his job
and requires a new occupation. He begins reading Shakespeare, Willa
Cather, Emerson, Thoreau, Robert Frost, Adrienne Rich. He joins a
book club. The readers decide to each bring in their own favorite story
and to read the first paragraph out loud. My father brings in a story I
have published, and he reads the first paragraph. They urge him to read
on, so he reads the second paragraph. They urge him to read further on
until he reads to the end.

Years go by. I am a writer; no one can deny it now. I am thinking
about a new relationship, perhaps a man with principles somewhat less
ferocious than the man I was last involved with. I am entering a new
country, the Land of Laziness. I return home after teaching and try
putting up my feet. I play old music—Little Richard, Tommy Jarrell,
Bob Marley and the Wailers. I putter among old snapshots, that terrible record of past times, and I throw them, a few at a time, into the

wastebasket. I paint pictures of the past in my writing and the worse the past gets the more the writing pleases me.

My father has just turned seventy-eight. He is the sole caretaker of my disabled mother. She, the genius of the family, the one who burned with ambition for her children to be educated, can no longer move. She can barely speak. And she won't be cared for by anyone but my father. My father tells me on the telephone that he works all the time, that there is never enough time to get everything done. I tell him I plan a lazy day, a Saturday of Sloth. I know we have come a long way when my father tells me that to him this sounds like a good idea.

4

Goodbye, Goodbye

Dr. Barbara H. Long (1923–2003)

Mother lies on the hospital bed, a slight form covered to the neck with a white sheet. Her head is massive, bony, white haired. Her cheeks are flushed as if she'd been out in the snow. Her brow is the color of plum petals, pale and only slightly creased. Her eyes slit and her rosy mouth puckers downward and makes a continuous slurping sound. She can blink, open her mouth, and, with great difficulty, swallow. Otherwise, she cannot move. Her knees are stuck bent, her foot bones twist, her arms stick straight down. Her fingers curl into fists, fingernails wounding palms. Her once stout body is tiny, diapered. She weighs eighty-five pounds. We cringe when the nurses call her "sweetie." She is a Professor Emerita of Psychology, author of dozens of research papers, a valedictorian, Summa Cum Laude PhD. Now she can say, "Help me" and "Okay" and "No."

She has the IQ of an Einstein. My twin sister and I agree: she should have been a Nobel Laureate. Instead, she married my father, a young dairy farmer, at age eighteen, and had her first baby that year. Ten months after my brother was born, she had twins, my sister and me. She was then nineteen. This was in 1943.

For ten years she lay in tangled sheets reading Agatha Christie novels. Or so I recall. From this never-made double bed, she commanded the household, broadcasting orders to the children, who eventually numbered five. She and my father grew up during the Great Depression and in no way benefited from the postwar boom. They were poor, without funds, with child. They shared a fervent belief in paying their own way. They opposed debt and they opposed television. They argued and shouted over whether to pay the gas bill or the phone bill or the doctor bill. They required their children to do well in school.

My mother disliked animals. We had cats, dogs, calves, cows, sheep, a pig, and a horse. We had a gaggle of nasty white-feathered geese. The rest of us entertained ourselves by picking up six-foot-long black snakes, writhing but harmless. We rode our horse and we rode our neighbor's horses. We children helped our father, the dairyman on a large commercial dairy farm. My father's boss paid us kids twenty-five cents an hour to help bring in hay, feed calves, scoop grain, shovel manure. We also rowed our rowboat out into the creek and into the Chester River. We ran around in the woods and made hideouts and forts and conducted warfare.

Dairy farming, its labors, practices, and decisions; its life in close proximity to animals; its very countryside held little interest for my mother. She lay in her bed reading, and every afternoon around five o'clock, she climbed out of bed to make dinner: "Get me a pie pan! Set the table! Get out the margarine! Pour the milk! Get Daddy! Diiiiiiiiiiiinner!"

We ate dinner in silence.

Every Saturday she baked for the week—apple pan dowdy (we called it apple pan downy), peach cobbler, Tizzy Lish Cake, donuts deep-fried in boiling fat. Every summer she supervised the garden—planting, weeding, picking. She commanded the process of putting up the winter's fare: freezing string beans and corn; canning tomatoes, peaches, and apples. We grew all our food except sugar, flour, macaroni, peanut butter, margarine, lard, and molasses. We raised our meat—each year a pig and a steer, out of which we got scrapple, roast cow meat, meatloaf.

We had no money. We did not have the price of a candy bar. We did not buy comic books or chewing gum. But my mother taught each of her children to read. Our house was filled with old books that had belonged to Walter Long, our father's father, and we children read every one of them, from *Wuthering Heights* to the Hardy Boys. When we were small, our parents read to us every day from the Bible. By the time we were six, they were reading to us *Pride and Prejudice* by Jane Austen (lost on us), and *Oliver Twist* by Charles Dickens (we loved that one).

We attended the white elementary school in Chestertown, Maryland, where until the fifth grade the teachers called each of us "twin." We could not learn arithmetic. On the other hand, we were the best readers in the class, possibly in the town.

One day—we Three Big Kids were ten or eleven—our parents called us together to announce that our mother would be going to college. Later she would say, "I stayed home with my children for ten years." She began to bathe every day, to wear girdles and dresses purchased from the Sears & Roebuck catalog. When in the afternoon she returned from her college classes she returned to her tangled sheets. But now, instead of reading mystery novels she read physics textbooks, calculus textbooks, Shakespeare.

She graduated Summa Cum Laude from college in 1957. Her image appears in the Metropolitan section of the *Baltimore Sun*. She is wearing a shirtwaist dress with a wide white collar. She does not look as heavy as I remember her. She is standing beside a tractor, and my father is on the tractor. The article reports with amazement that this mother of five children, this farmer's wife, is graduating Summa Cum Laude from Washington College.

She then went to graduate school in social psychology. She did not speak to us about her career. She did not speak to us about her struggles to become educated, if indeed she had struggles. She did not speak to us about any conflict between conventional motherhood and her motherhood. She did not speak to us about our clothes or our looks or our behaviors. In our opinion our clothes were deplorable. She had no interest in the matter. She did not ask us what we thought or how we felt.

She and my father took us camping to the Shenandoah National Park, where we pitched our tents every year under the same gigantic hickory tree. They took us to the Philadelphia Museum of Art or to Leary's Book Store on Market Street with its three stories, or maybe it was four stories, stuffed with books. They took us to classical music concerts. At one point—we were thirteen—our mother decided that her children should learn to jitterbug. She actually purchased a rock 'n' roll record, and our dusty living room was turned over to Bill Haley and His Comets playing "Rock Around the Clock."

Mother believed in Christmas, and every year the quarrels and stresses were put aside for the day. Bacon and eggs and biscuits replaced oatmeal, and we opened presents they'd spent untold efforts obtaining—one year, three new bicycles stood shining beside the Christmas tree. We were delirious with happiness.

When we were teenagers, when our mother was in graduate school, she researched private schools, and sent every one of us away to a private boarding school, combining scholarships and her graduate student stipend. No doubt she was the only graduate student in existence sending five children to boarding school. Meanwhile, my father continued as dairyman (his salary a pittance) on the farm.

When I was a senior in high school, I applied to Antioch College in Yellow Springs, Ohio. Antioch was more than my first choice: I desperately wanted to go there. But Antioch rejected me. My mother insisted that I write them a letter of protest, stating that Antioch was the perfect college for me, asking them to reconsider. I wrote the letter with my mother standing over me. Antioch replied with an acceptance. No doubt they had never before received a letter of protest from a rejected applicant. I was then seventeen years old. The one standing over me, the graduate student, my mother, was thirty-six years old.

She began publishing in graduate school, got her PhD, became a professor in the psychology department of a highly regarded college, eventually a full professor, eventually chair of her department. She authored or co-authored more than fifty research papers, and published them in journals such as *Journal of Social Psychology, Sociology of Education*, and *Psychological Reports*. The titles of her research papers reveal her life: "Parental Discord vs. Family Structure: Effects of Divorce on the Self-Esteem of Daughters" (my parents stayed married, kept quarreling); "Academic Expectancies of Black and White Teachers for Black and White First Graders" (my mother, an anti-racist resident of a segregated town, joined the American Association of University Women because it was the only integrated organization on the Eastern Shore of Maryland); "A Steady Boy Friend: a Step toward Resolution of the Intimacy Crisis for American College Women" (my mother wanted us to get married); "Attitudes toward Marriage among Unmarried Female Undergraduates" (were her daughters ever going to get married?).

Gifts from my mother. At Antioch I wore blue jeans and black turtlenecks, sandals in the summer, boots in the winter. Nevertheless, my mother sent a neon-pink full skirt with a matching neon-pink blouse. In later years gifts arrived wrapped in newspaper comics—ill-fitting polyester dresses, a Victorian pitcher made from greenware that she had painted by the number; a stuffed duck she'd made from a pattern; a poorly embroidered floral pattern scotch-taped to a small frame, these from the daughter of my Pennsylvania Dutch grandmother who was a skilled embroiderer. My mother, the genius, offered kitsch and schlock to her family of insufferable snobs. We were all, except for her, believers in fine things, traditional handicraft, high-quality workmanship, and (here excepting my father) even elegance, whatever that was.

We grew up, got married eventually, got divorced, got remarried. One of my mother's daughters, my younger sister Susanne, developed in her early thirties paranoid schizophrenia. My mother devoted days and weeks and years to helping Susanne fill out job applications, school applications, disability applications. She took care of her, instructed her, and encouraged her to try to live a normal life. A research psychologist, my mother commanded the literature on schizophrenia, spent a decade trying to save her child before Susanne committed suicide. At the loss of our beloved Susanne, my father sobbed into his hands. My mother sat in her familiar silence.

My mother adored her four grandchildren, became almost talkative around them, beamed at every new baby. And they adored her. She wrote them letters using vocabulary at their exact reading level. She monitored every grandchild's progress through school, devoted years of attention to the education of her less academically oriented grandson. She made sure he went to college and she made sure he graduated. It is my mother's legacy that our family is hyper-educated. Everyone (including, now, my father) has graduated from college. We have two PhDs, three master's degrees.

She never touched us or hugged us. She never asked us how things were at school. She never asked us how we were or how we felt. She never asked us how we were doing. She never spoke to our spouses. She didn't speak much at all. During family reunions she was mostly silent, but you would find her peering at you with piercing intensity.

Yet her brilliance was ever apparent. I never knew anyone else who could do what looked like casual but steady paging through a physics text at the end of which she had mastered the contents.

One of my favorite writers is Oliver Sacks, professor of clinical neurology, author of *The Man Who Mistook His Wife for a Hat, An Anthropologist on Mars*, and other books. Sacks's compassionate and appreciative accounts of people whose brains are differently wired gave me more tolerance for the somewhat strange brain who happened to be my mother. It comforted me deeply to imagine that she had Asperger's Syndrome, if only mildly. Asperger's is a type of autism combined with high and sometimes genius intelligence. The brains of persons so afflicted are physically different, particularly in the emotional center. People with Asperger's have an extremely weak body-sense and can't read body language, a severe social handicap. Thus, they might look away when speaking, or stare inappropriately, or leave the room in the middle of the other person's sentence. They are not among the best dressed because they have zero sense of dress. Sacks writes of the well-known autistic animal scientist Temple Grandin, "Temple sees her mind as lacking some of the 'subjectivity,' the inwardness, that others seem to have." She sees her disorder as "foremost a disorder of affect, of empathy." People with Asperger's may seem like emotional stones.

Many people think that Albert Einstein had Asperger's. He, contrary to popular perception, had poor social skills and weak interpersonal relationships. (See Andrea Gabor's *Einstein's Wife*.) Here's a story about Einstein. (I can't remember where I got it. Consider it hearsay.) He had a friend, the little boy who lived next door. Einstein went around with holes and tears in his clothes, and typically had bits of toilet paper stuck to his chin and jowls where he had nicked himself shaving with just water and a razor. One day the little boy, Einstein's friend, presented him with the gift of a tube of shaving cream. (Undoubtedly, the child's parents had put him up to this.) Einstein began shaving with the shaving cream. He was amazed! He was astounded! He could not get over what a smooth, nick-free shave he was getting. In the end, of course, he used up the shaving cream. What did he do? Did he go to the local drug store and purchase another supply of this easily obtainable, mass-produced product? No, he did not. He went back to using water, to nicking himself, to bleeding on himself, to going about with bits of toilet paper stuck to his face.

Now here is a story about my mother. One year my husband and I went home to my parents' house for Thanksgiving. The large family, now including spouses and grandchildren, sat down at the extended dining table with its tablecloth loaded with the traditional roast turkey, stuffing, cranberry sauce, candied sweet potatoes, creamed corn, rolls and butter, apple pies, mince pies. My mother sat at one end of the table, pouring milk. We passed her the tumblers and she would fill a tumbler and pass it along. The plastic pitcher she poured from had a rip going down its side, and with each pour, a stream of milk would land on the tablecloth and from there drip onto the floor. The milk at her place grew to a pond; the milk dripping off the table made a regular brook on the dining room floor. This did not perturb my mother. It did not appear to catch her attention. She kept pouring.

I thought: I will get my mother a beautiful new pitcher for Christmas! And so I did. A few months later, I again visited my parents and again we sat down to dinner. My mother began to pour the milk. I was amazed! She poured from the very same ripped plastic pitcher. Once again, the pond of milk, its efflux passing to the floor. "Mother!" I said. "Didn't I give you a pitcher for Christmas?????!!!!!!"

"Yes," she replied sweetly. "It's very nice. It's in the refrigerator with the apple juice."

Gradually I understood, even without a diagnosis, all her efforts to be a good mother, despite her aversion to touch, despite her difficulties in bonding to all but a very few. Gradually I saw the heroic effort she had made to succeed as a professional. Gradually I saw that her birthday gifts and Christmas gifts, which had once insulted me (a friend once quipped, "Has your mother ever met you?"), were attempts to give what she thought were pretty clothes.

My mother likes bright pink, bright red, and bright purple. She doesn't appreciate the difference between silk and plastic, doesn't speak to most people, beams at her grandchildren, writes intelligent letters to editors, continues to publish after she retired as a distinguished full professor. As she became increasingly disabled with diabetes and small strokes, as my father became her full-time caretaker, she focused more and more on her family.

For the past few years, we have talked on the telephone every Saturday morning. At last I feel close to her, connected. I have given up all resentment of this mother who was, in my adolescent opinion, abnormal, at times cruel, always cold. Now I feel nothing but compassion and empathy. I understand how at age eighteen she entered into a way of life for which she was poorly equipped; how by age twenty her babies plus 1940s female-role expectations carved in stone locked her into an implacable destiny; how she—without discussion, without a support group, without company, without question, without hesitation, without ambivalence, without social context, without a single friend— broke out of that prison and excelled in a distinguished career of which her family is the beneficiary.

Now her time is almost over. Mother lies in her hospital bed all skin and bone, her massive form shrunken, flesh reduced to skullbone, cheekbone, backbone. Her joints have locked straight or bent, rigid as rusted machine parts. She is mind and bone, pain and bone, voice and bone. Her voice flutters, tiny, slurred. I lean close, ask her what she wants, try to understand. "Do you want to turn toward the window?" I ask. "Say yes or no." Death is in the room, rattling his cart of bones. Mother ignores him. She says, distinctly, "Yes or no."

"Very funny, Mother," I say.

She can no longer laugh, but something in her face tells me she thinks it's funny too.

5

Inheritance

For, up there in heaven, isn't paradise an immense library?

GASTON BACHELARD

It was my mother who gave *me* writing, but who gave *us* writing? Once I would have blithely stated: the ancient Sumerians, people of the Euphrates River, of ancient Mesopotamia. And yes, the Sumerians wrote in cuneiform on clay tablets and yes, they did invent writing. It's just that they weren't the only inventors of writing. Sumerian tablets excavated from the libraries of ancient Mesopotamia list items of trade—beer, grain, wool, sheep. Centuries later, Sumerian scribes and bards scratched the world's first epic—the story of the hero Gilgamesh—into clay tablets. But there were other origins of writing. Egyptian hieroglyphs, completely uninfluenced by cuneiform (according to the historian of ancient scripts Marc Zender). Mayan writing. Chinese writing. And Linear A, the writing of Knossos on the island of Crete, which developed into Linear B, which developed into ancient Greek.

I first wrote with a stick on the edges of mud puddles. As a child I composed stories and inscribed them in mud before I knew my letters. With my stick and my mud-puddle compositions, I inhabited a world made by the first scribes.

We children were told Bible stories, and these were my first stories. Jesus, who suffered the little children to come unto him. The coat of many colors. The flood. Noah's creatures embarking the ark two by two.

George Smith's discovery in the 1870s of clay tablets inscribed in cuneiform with the Gilgamesh epic excited the world because the epic records a much older flood story. The Remote One relates to Gilgamesh the story of treacherous high water, how Ea, Lord of the Clear Eye, tells the people: "Build an ark. / Abandon riches. Seek life. / Scorn possessions. Hold onto life. / Load the seed of every living thing into your ark, / the boat that you will build" (Gilgamesh, Tablet XI, Column i, trans. by John Gardner and John Maier).

We inherit this flood. It seems that only a few lived to tell the tale, which is lying beside my computer in the form of John Gardner and John Maier's translation.

Another old book, lying across the room on my handwriting desk, came into my hands from the world of my mother's childhood. The blue buckram covers of *The Mary Frances Sewing Book: Adventures among the Thimble People* are threadbare. The cover illustration has faded from vivid orange to pale mustard. It shows little Mary Frances wearing her ruffled white pinafore among her assistants, the Thimble People. All through the yellowed pages, Mary Frances consorts with pincushion persons and a doll named Angie who is terribly cross.

Mary Frances is given to saying "Dear me!" She receives visitations from the Sewing Bird, who says, "Good Afternoon, Your Seamstress-ship!" Each chapter has a double-page folded into a pattern packet inside of which can be discovered a tissue-paper pattern for doll clothes—pinafore, workbag, muff, kimono, underwaist, morning dress, bloomers.

The Mary Frances Sewing Book provides thorough instruction for the apprentice seamstress, information on how to do the running stitch, the combination stitch, the overhand stitch, the hemstitch, the blanket stitch, and its sister, the buttonhole stitch. It lists kinds of needles and threads (No. 6 for coarse work; No. 9 for fine work), and instructs Her Seamstress-ship in how to finish lace sleeves and in how to make a fell. (A fell is a seam hemmed down to prevent edges from raveling.) Fairy Ladies and Sewing Birds arrive to convey the mysteries of their arcane craft with encouragements like "Thank you very much dear child!"

As a child, I loved *The Mary Frances Sewing Book*, even though I never learned to sew from it. Now it connects me to my own childhood, to my mother's childhood in 1920s Bucks County, and to the world of my Pennsylvania Dutch grandmother, Olive Erisman Henry, an expert seamstress.

It connects me to the vanished world in which women manufactured clothing, in which every girl was taught to sew. No doubt it's a lucky thing that this world of girls learning to sew has been superseded, that now a girl can practice chemistry or rocket science. But little Mary Frances, in these old pages with their orange and black illustrations of smiling workbaskets and chattering thimbles has more skill at smocking and buttonholing and facing and hemming than all but a very few in the world today. It was a high, intricate skill requiring an elaborate apprenticeship, worthy of respect.

Holding *The Mary Frances Sewing Book* or *The Gilgamesh Epic* in my two hands, I hold Exhibit A of the history of the book. You can turn book pages frontward or backward at your pleasure, and you can mark your place. You can line books up, spine out, and read their labels on their spines.

The book, that convenient container for the written word, did not of course always exist. In Egypt writing was done with a reed dipped in ink on long rolls of papyrus. Papyrus was made from the sedge *Cyperus papyrus* abundantly available on the Nile River. The reader unrolled the roll as he read page-sized columns in one direction—the British Museum has a papyrus text that unrolls to 133 feet. Unfortunately, papyrus disintegrates in damp climates within about a hundred years. Thereby have most of the literature and records of the ancient world been irretrievably lost. Sophocles wrote 113 plays, of which 7 survive.

After papyrus came vellum (calfskin), goatskin, and other tanned skins. Scribes copying texts far from the Nile considered these writing surfaces inferior to papyrus, but they made do with the material at hand. Their writings have survived.

Someone thought of folding papyrus between the columns so the thing could be flipped back and forth like an accordion. Put between oak boards, this was called a codex, Latin for book. Eventually the folds of codices were sewn at the back and cut in the front. Labels were put on spines. Then came Gutenberg.

Books clot my small house, stacked to the ceiling in bookcases, scattered on tables, arranged in some meticulous but forgotten order on the carpet along the foot of the bookcase, teetering on the coffee table, hiding under the bed . . .

I like the ancient texts. Sappho's poems and Sophocles's plays, remanufactured from their papyrus originals, have their place in my library, ready to be replayed like musical compositions. So also the Rig Veda; the I Ching; the Bible; Homer; Ovid; the works of Hildegard—that medieval illuminator, scholar, and mystic. And the poets. John Milton and May Swenson, Rilke, Rukeyser, et al. extend to the very edge of my self-imposed but strict six-foot limit on poetry books. Novels, each chunky one replete with its promise of a long read on a lazy winter Saturday, a fire snapping in the fireplace, bread rising, Bach floating out of the music player.

A book can stain an entire era with its peculiar coloration. A book's thoughts, terrors, transformations, trepidations, and fated plot-twists can mirror our own or become our own.

I am twenty years old. I sit in a small room lined with gray bottles, reading a book. A small window admits the light of a gray sky that turns the near bottles silvery gray. It is evening, the swing shift at Doctors Hospital in New York City. This is the transfusion dispensary. I have been provided with a soft-seated office chair to read in, and a small Formica-gray counter. My companions are the gray bottles rising from tiled floor to florescent-lighted ceiling. It is four hours into the eight-hour shift, and as yet no patient has required a transfusion and no nurse has appeared at the Dutch door with a prescription for one of the gray bottles. I am reading the novel *Demian* by Hermann Hesse. I am the same age as the narrator, who sits on the bottom step of his parents' house and "abandons himself to misery."

I too abandon myself to misery. I too am "haunted by misfortune." My boyfriend is home at our apartment, or maybe not. He is at the deli or having a party, or a good talk, or a good dinner, or whatever he does during the long hours from 4 in the afternoon until 12 midnight, while I sit with my gray bottles at the Doctors' Hospital, reading, reading,

reading, sinking into the miseries of another young person in another time in another city. The gray bottles keep their own counsel.

I arrive home after midnight to find my boyfriend and our hip and pretty roommate deep in conversation, discussing world politics, smoking cigarettes, talking excitedly, sipping wine. Although they have ostensibly waited up for me, they barely notice my entrance. I go to bed. I wait for my boyfriend, fall into a restless sleep, disturbed by the talk in the kitchen that winds on and on like a long dream.

Day after day, I set out for work just before my boyfriend is to arrive home. I arrive at the hospital and go to the gray room and sit with the gray bottles. I read page after page. I am sinking. I am sinking into the tortures and miseries and terrible anxieties of our hero. I am becoming suicidal.

Finally, I quit my job. I quit my boyfriend. I abandon our hero and his lonely reveries to the gray counter in the gray room with the gray bottles. Some things—life—are more important than a book.

I moved to Seattle in 1988 in part because I felt a kinship with this book-obsessed city. Everywhere you look, you see people engrossed in a book—on the Metro, at the lunch counter, at the library, even on the curb, under the bridge. In Seattle, the homeless read, and the home-bodies also read. Recently, at a reception following a poetry reading, I mentioned my little six-foot regulation—I may own a maximum of six feet of poetry books. The response was one of amazement. Six feet seemed unduly harsh, a case of penury if not self-flagellation.

In Seattle, bookstores survive, even if they don't thrive in this new digital age. Every Seattle neighborhood has one or more coffeehouses where people are reading and writing. People are reading Kant or Ursula K. Le Guin or *The Hobbit* or Richard Wright once again. Or they are reading Colette. The fabulous Colette.

How she flung her books from her generation into mine, like an Olympic discus thrower, and I have stumbled after, emulating, trying to match her moves as I have done my whole life to the writers I love, because the writers I want to read are the hot ones, hot for life, hot for love, hot on the page, desirous of experiencing everything, and they never skim or say a conventional thing but churn churn churn like a boiling white-water river rushing through a forest, and in it you see sky and earth and passion and grief, and even silence, even moments of pure ineffable grace and you throw yourself into the river and let yourself be swept away, the waterfall approaching, loud in your ear.

So I read Colette. So I burn a little with all she felt and all she put down, sitting at her elaborately accoutered desk, concentrating with ferocious concentration until she got colder and colder, writing with rugs on her knees, writing sheet after sheet, writing with Proust in her ear, for Colette reread all of Proust every other year. Colette lived so intensely, according to her husband, that she could not walk into a garden without touching everything, separating sepals of flowers, licking poison berries and leaves, scratching bees on their backs, and she would emerge littered with bugs in her hair, coated with dirt, twigs, petals, pine needles.

Reading Colette, I wander in her garden, I buzz with her bumblebees. I feel her obsessive loves, her sadness, her boredom. Writing with my stick in a Pennsylvania mud puddle, I am her child. I grow up to read her words, to enter into her secret garden. I grow up to be a writer. I step in her footprints, in all of their footprints. I try to write like them. I try to write different from them. They're all looking after me, I'm pretty sure. They're chiding me and egging me on. They're kicking my butt and patting my back. Sometimes they give up on me. But they always come back. As well they should, for I am their inheritor.

II Fire

A winding stair, a chamber arched with stone,

A grey stone fireplace with an open hearth,

A candle and written page.

YEATS, "My House"

6

Throwing Stones

It is our fate and misfortune that we live in history.
There is nothing we can do about it. We should learn to swim in history.
VITALY KOMAR AND ALEX MELAMID

I once had visions of becoming one of the Great Minds of the West. This was in the late 1950s at Moravian Seminary for Girls in Bethlehem, Pennsylvania. At Moravian Seminary we wore uniforms with white anklets and saddle shoes. We attended chapel every morning, and we also prayed before breakfast, dinner, and study hall. In our eleventh-grade history class, taught by Miss Fanny Costello, we studied the French Revolution.

It was exciting becoming one of the Great Minds of the West. I would go to the library, an oak-paneled room lined with books and furnished with refectory tables. I would spend an hour puzzling through two or three paragraphs of Hegel in one of the *Great Books of the Western World*. I intended to read every Great Book of the Western World. The librarian, Miss Hartman, had never known another Moravian Seminary girl to do such a thing, and she liked me. I was drawn to thick books. I would read them and keep score, rather like counting laps in a swimming competition. At seventeen I had long since read *War and Peace* (1,483 pages), *Les Misérables* (1,463 pages), *The Hunchback of Notre Dame* (688 pages), and *The Three Musketeers* (555 pages). Miss Hartman encouraged me to read books based on factors other than number of pages. I read every book she suggested, including *The Bridge over the River Kwai*, *Archy and Mehitabel*, and *Till We Have Faces*.

In our history class, the unit on the French Revolution was especially exciting. I scorned our textbook; possibly I didn't read it. With happy anticipation, I tackled both Edmund Burke's *Reflections on the Revolution in France* and Thomas Carlyle's three-volume *The French Revolution: A History*. Carlyle was thrilling! Of some desperate finance minister he wrote: "What could a poor Minister do? . . . A sinking pilot will fling out all things, his very biscuit-bags, lead, log, compass and quadrant, before flinging out himself. It is on this principle, of sinking, and the incipient delirium of despair, that we explain likewise the almost miraculous 'invitation to thinkers.'" And so on.

Already I had made July 14—the day the Mob liberated the Bastille—my own personal holiday. I made Liberté, Égalité, Fraternité my own personal motto. I fervently believed every word of the Declaration of the Rights of Man. I could almost feel the cold blade of the guillotine on the hot neck of Marie Antoinette.

I arrived at class each day full of excitement, and often raised my hand to offer a comment or to answer a question. I was Miss Costello's favorite eleventh-grade girl, I was pretty sure.

I fancied myself a thinker.

The day of the test arrived. I entered the old wood-floored classroom filled with a sense of expectation and impending triumph. On the test I wrote and wrote, pouring myself forth on the French Revolution. I finished with a feeling of pride, imagining how pleased and possibly even excited Miss Costello would be.

The next day we were to get back our tests. We entered the classroom and took our seats, fifteen girls wearing pastel uniforms with round white collars and white anklets and white-and-brown saddle shoes. Miss Costello entered. The girls stood. Miss Costello said, "You may be seated." The girls sat and folded their hands on their desks. Miss Costello put her briefcase on the desk and drew out the tests. The class waited in silent anticipation. Without a word, Miss Costello went from desk to desk, returning the tests.

I looked at my test. I could not believe my eyes. On the test Miss Costello had drawn a large red F. I was in shock. I kept looking at the F to make sure it was really an F. I turned the test over to see if there was a mistake. Then I turned it back to the front. Miss Costello had written a single comment next to the F: How can you write about the French Revolution without mentioning Voltaire? Voltaire. I had forgotten about Voltaire.

Miss Costello was speaking to the class. She was a small-boned woman with straight black-and-gray hair, which she parted on one side and barretted on the other. She wore a sweater set and a pleated plaid skirt and stockings and polished brown loafers. She had a thin mouth and pale papery skin and she wore granny glasses and idolized Arnold Toynbee. Now she was talking on and on about something, but I did not hear her. When the bell rang, I rose and trooped out of the classroom along with the other girls. I did not look at Miss Costello, nor did I speak to her.

I stopped reading books on the French Revolution. Instead, I memorized the textbook. The next test was a True/False test and I answered all the questions correctly. I finished the questions twenty minutes before the end of the class and sat among the busy test-takers with my hands folded on my desk, in scornful silence. On that test, I got an A.

So that proved it. The conventional ones, the little minds, the obedient ones, the ones that studied for grades and learned by rote, the ones that couldn't care less about the French Revolution, or about Hegel or about Tolstoy, they were the ones who got the stamp of approval. I proved to Miss Costello that I could do that too, if I wanted to. But it was beneath me, and it was too late for Miss Costello. I never spoke to her again.

I graduated from Moravian Seminary and went to college and shortly after that I entered into the shadowy realm of American rebellion, into the sixties of pickets and protests and street marches and flag burnings, and I wore blue jeans and black turtlenecks and sandals and grew my hair long and smoked dope and read Gregory Corso and Allen Ginsberg and played bongo drums and danced all night and marched against the war and read Gramsci and Marx and Simone de Beauvoir and Virginia Woolf and hawked newspapers on street corners and waited on tables and learned how to throw stones at the plate glass windows of American banks and how to laugh at police officers and dodge their lobs of tear gas and yell Oink! at the top of my lungs.

7

Autobiography

An A–Z

There is rust in my mouth,
the stain of an old kiss.
ANNE SEXTON

Is autobiography a fiction, a delusion, a defense to the jury? To what extent does it reflect our era, our notions of what *ought* to be, our conventions, our common assumptions whether conscious or unconscious? To what extent does it deviate from what *did* happen? Does our autobiography shift as we age, becoming more sanguine if we are happy, more despairing if we are unhappy? Whatever the case, here's my stab at my own autobiography. It's ordered by way of the alphabet, for what better way to set a limit on the random and continuous movie that is memory.

Arrests. Gwynn Oak Amusement Park. Baltimore, Maryland. July 4, 1963. We are black and white together and we shall not be moved. At the locked gate, screeching hecklers, cops everywhere. White faces bleed hate. Our group announces, "Let's go back to the bus." We walk back toward the bus along a narrow creek lined with trees, the boundary of the amusement park. No fence. Voila! We cross the creek. We find ourselves in a flat grassy field, hot sun, carnival rides in the distance. We are shocked, without plan. We had not expected to get into the park. We say, "Let's go to the merry-go-round." We link arms and inch forward, one multilegged creature. White thugs materialize. They surround us, take off their belts. Shouts—distant—from the direction of the entrance gate. Our arms link tight. If one falls, we're all down. We begin singing "O Say, Can You See." Fear cracks our voices. But these racist thugs are patriotic. They won't start beating us till we stop singing the National Anthem. We don't stop singing. We sing and we sing. Will the cops ever come? Do they even know we are here? At last they arrive. We sink to the ground. In the paddy wagon we sing at the top of our lungs, "And before I'll be a slave, I'll be buried in my grave, and go home to my Lord, and be free."

Bottle-fed baby. I'm an identical twin, the third child born in ten months to my teenaged parents. "One time," my mother said, laughing, "Pammy got two bottles and you didn't get any!"

Coal. Coal is the rock that burns. But what burned in me, what engine drove me to twenty years of research into the American coal-fields, carried out after work and on vacation? Was it the memory of Pennsylvania anthracite rattling down the chute into Gran's Bucks County, Pennsylvania, cellar bin?

Dithering. How much time do I spend dithering? Plowing through e-mail. Going to Facebook to read about nothing. Surfing the Internet to see if I can track down an old lover I haven't seen for forty years. Reading a chapter in a book and then reading a different chapter in a different book. Sweeping the floor. Washing dishes. Sorting through piles of photos, wondering which baby this is, whether I should keep this snapshot, wondering what will become of these hundreds of snapshots when I'm gone.

Elvis. Elvis was king. And after Elvis, Chubby Checker, doin' the twist. Then the Beatles, the Rolling Stones, the Mamas and the Papas. Back to Little Richard, the one and only. And after Little Richard, Jerry Lee, also the one and only. And after Jerry Lee Lewis, the great Appalachian fiddler Tommy Jarrell. Then Dolly Parton—composer of three thousand songs. The immortal Otis Redding. The immortal Bob Marley.

Back to Bill Monroe. Back to Ralph Stanley. On to Jim Morrison. On to Thelonious Monk, Miles Davis, Coltrane. Back to Ben Webster.

First love. He played guitar. He sang "Laughin', Free, and Gone" and "Oh, Miss Mary." He talked politics long into the night. He was a student leader in the Student Peace Union. Thick kissable lips. He dropped out of school, moved back to New York, got into computers. He wrote me disquisitions about the test ban treaty, about the Cuban Missile Crisis, about the flotilla of traffic winding below his night window, about getting laid and sharing a smoke after getting laid. I dropped out of school and went to New York. We shacked up. Then we broke up.

Getting up to write. I get up to write. I write in my journal. I write to wake up. I write to drink espresso by. This autumn I am writing in Journal No. 304. I write about nothing. But nothing can come to something just as something can come to nothing. Every morning I draw letters with my fountain pen. I make letters make words. Words about nothing.

Hero. Peter refused to go to Vietnam. He went instead to jail. As soon as he got out, we got married. He went to graduate school, got a PhD, went to law school, got a JD. He began writing books. There were the Boston years, the San Diego years. After twenty years, we got a divorce. After that we talked on the phone every day for five years. Our friends praised our divorce as if we'd remodeled our house or won the lottery. Now we've been divorced for as long as we were married. Now we're just old friends.

I. Who am I? Who am I when I'm sleeping? Who am I when I'm dreaming? Am I still the third child born to my parents during the war year of 1943? Am I still a reader? Am I a writer when I'm sleeping? Am I a twin? Am I a Seattleite? Am I still descended from Scottish and English and Pennsylvania Dutch immigrants? Am I still 2.9 percent Neandertal? When I am sleeping, what happens to my opinions? Am I still a woman? Am I anybody?

Jay. Jay wears red and yellow shirts printed with birds or flowers, as if he were on vacation in Hawaii. Jay loves his jalopy. Jay carries a water bottle and wears an earpiece to talk on his cell phone without getting brain cancer. Jay writes a haiku every day. Jay writes books about spiritual seeking. Jay jogs. Jay goes to jazz-dance class and Jay dances.

Kids. We "Three Big Kids" built forts, went swimming in the Chester River, read books, rowed our rowboat, rode our bikes, played Cowboys and Indians, ignored Susanne, played Monopoly.

Learning. What am I learning today? I'm learning the plants. I am learning a few of the 310 species of trees growing in Seattle. I am learning the cascara tree, its roundish leaf with a little point, its bark harvested, traditionally, for use as a laxative. I'm learning serviceberry and huckleberry and the cherries—sweet cherry and bitter cherry and cherry plum and chokecherry. I'm learning the conifers—Doug fir, grand fir, western hemlock, western red cedar, redwood, sequoia. I'm a slow learner, but I'm learning.

Music. That year, 1973, I practiced my clawhammer banjo, sold records at bluegrass festivals in the Deep South, filled orders at Rounder Records, marched against the war, sang "Trouble in Mind" and "Johnny I Hardly Knew Ye" and "House of the Rising Sun," read "Kaddish" by Ginsberg, posed for art classes, wrote poems, made love to Jony.

Narcissism. I am staring into the pool to see if I can find my own rippling image. I see nothing but minnows. Are these minnows me? Am I minnows mixed with water and stones? Do I have minnows in mind? Is my mind a matter of minnows?

Oneiric Autobiography. What if we each wrote an autobiography consisting only of our dreams?

Printer. You lift cartons of stock, cut the stock, set up the press, and run the press. You bring the stock down the steep basement staircase. You cut the stock, set up the press, run the press. You set up for envelopes and run the press. You set up for a two-color job and run the press. You've got your loupe. You pull a sheet, hold it up, peer into the loupe to check if the ink is lying down. "It's lying down good, no emulsification." You pull another sheet, check the registration. The owner hired you out of desperation for more labor. He was a kindly man who believed women should cook and clean. He was surprised that you could print. He was surprised that you could bring the paper down. He was surprised that you could run your hot-rod press fast. Vroom!

Quantities. Height, 5 feet 4.5 inches. Weight, 143 pounds. Amount of water used per day, 48 gallons. Amount of electricity consumed per day, 4.46 kilowatts. Amount of natural gas consumed per day, 1.66 therms. Amount of caffeine consumed per month, 65 mugs. Number of words written per day, 500. Number of TV programs watched per day, 0.

Rape. Every Monday evening for two years I attend the PTSD group at the Harborview Center for Sexual Assault and Traumatic Stress. This is in Seattle. One week our kindly and perceptive therapist asks us what we think of the idea of men therapists sitting in on the group. Alma says, "Do whatever you want, but I will just lie." I say, "It's fine with me as long as they're not rapists." The therapist looks a bit startled. "Oh," she says. "No. They wouldn't be rapists. Definitely not." The subject never comes up again.

Susanne. Susanne was an ethereal and rather stunning beauty. She was funny and creative. She joined the Peace Corps. She took photographs, made homemade yogurt, tie-dyed T-shirts, played the recorder, knit scarves and sweaters, painted the sumi-e way, made watercolors, and taught English as a Second Language. She loved her students and her students loved her. And we loved her. And she committed suicide.

Twins. We twins rode our old draft horse, went out to get the cows, fed the calves, washed the milking machines, did the ironing, made the school lunches, walked down the mile-long dirt lane to the school bus, rode the bus to Chestertown Elementary School, rode the bus home, walked back up the dirt lane, fed the calves, did our homework, ate raisins, hated arithmetic, played with our dolls, read *Cherry Ames, Student Nurse,* went barefoot, brushed our 4-h calves, mowed the front yard, read books, read the Bible, drove the tractor during haying time, weeded the garden, picked string beans, helped Mummy freeze string beans, picked peaches, helped Mummy can peaches, put out the garbage, practiced public speaking for the 4-h club, refused to dress alike, climbed up and down the wooden fire escape to our attic room, whispered to each other, "gave each other confidence," picked strawberries, learned to make change, sold strawberries door to door in town, stayed away from Daddy's bees, helped put labels on the honey jars, sold honey door to door in town, wrote letters to Grandma, watched our brother beat up Susanne, ignored Susanne, collected plastic elephants, collected stones, won third and fourth prizes at the county fair, prided ourselves on not being city slickers, read *How to Be Tops in Your Teens,* took a bath once a week, went to Sunday school, went to church, did the wash, hung out the wash, did the ironing, read every old book in the old house, went to the library, read *The Power of Positive Thinking,* complimented the string beans, complimented the cow roast, complimented the potatoes, drank milk, complimented the milk, helped Daddy milk the cows, learned to do the Fox Trot, waited to grow up, waited to grow up, waited to grow up, waited to grow up.

Universe. The universe is about 13.82 billion years old. Earth is about 4.6 billion years old. Life on Earth is about 3.8 billion years old. I am, at this writing, seventy-three years old. I illustrate life on Earth. I illustrate the evolution of life forms into species. I'm animal, chordate, mammal, primate. Genus, *Homo.* Species, *Homo sapiens* salted with 2.9 percent *Homo Neanderthensis.* I'm a *Homo sapiens sapiens* (species specializing in making abecedarians).

Vietnam. We tape a map of Vietnam over the gaping hole in the door. We are Antioch college students on our coop jobs. This door leads to a tenement flat on the Lower East Side. It is 1965, the year of the Body

Count. Then came the body bags. Hell No, We Won't Go. Or did go. Or went, but did not return. Or did return, but different, all fucked up, forever changed.

Winslow. Edward Winslow was born in Droitwich, England, in 1560. Edward Winslow married Magdalene Ollyver of London. Their son, Kenelm Winslow, was born in 1599. He arrived in Plymouth, Massachusetts, sometime before 1633, and married widow Jane Eleanor Adams. Their son, Kenelm Winslow 2nd, was born in Plymouth, Massachusetts, in 1635. Kenelm Winslow 2nd married Mercy Wordon, and their son, Kenelm Winslow 3rd, was born in 1668 at Situate, Massachusetts. Kenelm Winslow 3rd married Berthia Hall. Their son, Kenelm Winslow 4th, was born in 1700 in Harwich, Massachusetts. Kenelm Winslow 4th married Zerviah(?), and their son, Deacon Nathan Winslow, was born in 1737. He became a farmer in Harwich, Massachusetts, and married Eunice Mayo. Their son, Seth Winslow, was born in 1764 in Harwich, Massachusetts. Seth Winslow married Hannah Crosby, and their son, Seth E. Winslow, was born in 1790. Seth E. Winslow became a medical doctor. He married Sarah Giles. Their son, Stephen Noyes Winslow, was born in 1826. Stephen Noyes Winslow became "the grand old man of journalism" in the city of Philadelphia. He was my great-great-grandfather. He married Sara Breesh, and their daughter, Clara Elizabeth Winslow, was born in 1858. She married Howard Long, and they had a son, Walter Long, born in 1883. Walter Long became a writer. He married a Scottish widow, Annie McIwrick Sproul, and their son, Winslow Long, was born in 1922, in Liverpool, England. Winslow Long became a farmer. He became my father.

Xenium. This abecedarian is a xenium, a gift to myself. An attempt to discover who or what I am, really. May it also be a xenium (ZEEnium, gift) to Scrabble players. The word "xenium" went obsolete back in the 1930s but deserves to be reunited with our little cupful of words starting with x. Memo to the editors of *Oxford English Dictionary*. Xenium is hereby back in use.

Yoga. I go to my yoga class. I twist and I bend. I do cat/cow, the bridge pose, downward-facing dog. I stand up, perform a deep bow. To what do I bow? I bow to the muse, low and deep, back straight, hair to shins. And then I do the warrior pose.

Zen. Buddhism is the road I travel toward. But never arrive. I'm more a pseudo-Buddhist. I don't believe the right things. I don't meditate the right way. Or maybe I worship the muse in the church of poetry. Which maybe is the path of a true pseudo-Buddhist. But I'm more Buddhist than Methodist, more broad-minded than bigoted. I've left to my childhood those pink-skinned bigots who sang like crows, those small-town small minds, those segregationists.

8

Banjo

Six Tunes for Old Time's Sake

Who were we, and why did we live?

CAROLE MASO

In the early 1970s, before I became a printer, I became intensely involved with traditional American music. I practiced my banjo every day and even got pretty good at it. If you know old-time music, you know that traditional banjo tunes—the old fiddle tunes—have two parts, a low part (Part A) and a high part (Part B). So do the tunes I present here each have two parts.

The tunes are dedicated to Danny, an old friend who disappeared from my life without a trace. I miss him and wrote these lyrics for him.

Shady Grove

Once I had an old banjo.
The strings were made of twine.

A.
My banjo had gut strings. A skin head. Danny, you may remember it. Not a crass, high-wired chrome-plated amphetamine-pumped flashy badass bluegrass banjo. Once I had an old banjo, a clawhammer banjo with a skin head the size of a shoofly pie. The only tune that it could play was trouble on my mind. Oh—and "Cripple Creek." It was a plunk and knock banjo, a clucker that played "Cluck Ole Hen" pretty good. It sounded more like a branch tapping a window in a storm than any

more downtown sort of sound. It could be, Danny, that you will remember a different banjo. It could be that this banjo I am remembering is a cross between the banjo I once had and the banjo I once wanted. Keep in mind, old friend, that memoir is suspect. Memoir is stories patched together like a patchwork quilt. Memory stitched with artistry. We like it to look pretty good—wherever we get our scraps from.

B.

The banjo I am thinking of stands for an era—1973 or 1974, Somerville, Massachusetts. You and I were both hangers-on or hangers-in at Rounder Records, me packing records in Somerville and selling records at music festivals for twenty-five-dollars-a-week-plus-peanut-butter-sandwiches, and you doing I don't remember what. We were lovers, but friends more than lovers. You were a lot younger than I was, easygoing and easy in so many ways. Still, there were things we never spoke about.

Childhood, for one. On the Eastern Shore of Maryland there's an old slave town on the old Chester River where in the 1950s you could stand outside the brick jailhouse and hear black men pluck banjo tunes.

Let that piece of American history be part of this tune.

Those days, it was my job to iron the school clothes, which I did listening to hillbilly radio. The aspiring longhairs in the Long family would poke fun at me listening to "Your Cheatin' Heart" and "Lovesick Blues" while I ironed sleeves and cuffs and ruffles. Looking back on it, figuring out the dates, it's very likely I was listening to Hank Williams himself, live on the Grand Ole Opry, at the height of his fame just before he died in a car crash on New Year's Day 1953. I was ten years old. You, Danny, were just a newborn. Hank Williams was twenty-nine years old. We were all pretty young at the time.

Hank Williams learned his music from a black street musician named Rufus Payne, known as Tee Tot. On the street in Georgiana, Alabama, Hank Williams learned to sing and he learned to drink whiskey.

The Drunken Hiccups

> If the ocean was whiskey and I was a duck,
> I'd sink to the bottom and never come up.

A.

Whiskey turns the world to gold. Whisky puts a glow on the polished wood of the bar, on dim bar lights and glinting bottles, soft-focused like memory or like sepia-toned nineteenth-century photographs. Whiskey turns the world into an old movie, nostalgic and mildly romantic. The band is setting up to play. The barroom is crowded, glowing with good cheer. Mildly inebriated now. Pleasantly drunk now. No one yet barfing at the curb. No one yet spinning down like a top to break a tooth on the broken pavement. The band starts pouring bluegrass into the barroom like an intoxicating wine, and they lift the acolytes, groupies, wannabes, and drunks into a swoon. Tonight even the band gets drunk and the lead singer croaks.

The morning after, the bar is barren, crude, stale, dingy-walled. Beer pools in little puddles on the floor. Bar stools barf stuffing. The transported, transfixed, cheering audience in all its various parts is waking up hung over, a few with some man or woman whose name has slipped away in the night, and there is a Sunday to get through and a Monday to get through, and music is a thing of the past, like happiness.

B.

One Sunday evening I take the night off and walk to Harvard Square to browse in a bookstore and go to a movie. I love going to the movies by myself. I sit up front, place myself directly within that bright altered world, not a spectator, not a voyeur, but a participant. I stay until the last credit rolls.

On this evening, I get to Harvard Square and run into a friend, a pal younger than myself, really the friend of a friend though we've been in each other's company many times. This person—we might as well call him Tobias since I don't remember his name—confides that he is in possession of some excellent weed. He wonders if I would like to partake.

Yes I would. We withdraw to a park bench secluded by a grove of rhododendrons provided by Harvard University. Tobias gets out his tobacco pouch, removes from it a plastic baggie of dried leaves and stems, and a package of cigarette papers. He rolls a fat marijuana cigarette, takes a deep toke, and hands it to me. I take my toke and we pass it back and forth, meanwhile chatting about his business buying and selling sundries and antiques, working out of a cloth spread on the pavement of various parks and public places. When it's time for the movie to start, Tobias walks me to the Harvard Square Theater and we cheerfully say goodbye.

I pay for my ticket and enter the dark theater and take a seat second row from the front, three seats in from the aisle. The previews run and then the movie begins. I sit in the dark, gradually feeling heavier and heavier, as if my body had surreptitiously gained three hundred pounds or had become, at some point when I was not paying attention, a building. I am completely immobilized by my new bulk and weight. I cannot move a finger or a toe.

At the same time I float to the ceiling. I understand distinctly, but as if observing someone else, that this is a social emergency. If I wait until the end of the movie to get up, the chances are excellent that I will be unable to move, followed by an embarrassing scene with an usher or janitor, *as it were*.

I begin inserting the phrase *as it were* into each of my thoughts. It is necessary, *as it were*, to rise and leave the theater at once, because if not now, *as it were*, never. I may be stoned ("turned to stone"), but I do not like to lose face. I gather my shoulder bag. This is an entirely mental action, *as it were*, because I am clutching the leather strap in my hand, which has not moved. With a violent effort I stand up, move out to the aisle, and float, *as it were*, out of the theater, looking straight ahead, never minding the stares of the popcorn persons.

Out on the street I can walk pretty good, *as it were*, but my already weak ability to gauge distance has vacated. I float to the curb with the Sunday night crowd and wait with them for the light to change. I cannot tell which lights attach to which cars, so I am grateful, *as it were*, for the unwitting assistance of the crowd crossing the street. I cross with them and float along in the direction of my house, about a mile away, *as it were.*

The lights are very pretty. I develop the excellent, *as it were,* technique of crossing streets by keeping to the ranks of the other street crossers. I float along, flapping my feet. But then I start getting to street corners with no crossers. I wait there for a companion of the crossover to arrive, then I cross with the person crossing. To appear to have a reason to be just standing there, I fumble in my shoulder bag for something. This works very well, *as it were,* but the closer I get to home, the longer I have to fumble in my bag until somebody appears. Finally, I arrive at the busy street—Somerville Avenue—right across from my house. Cars are whizzing by. I wait there for what may be an hour. I am just very patient. I just know somebody will come. Finally somebody does come. Together we cross the street.

I am home. I let myself in, mount the stairs to my second floor apartment. There is my kitchen with its table and yellow tablecloth, with its cast-iron skillet hanging on its own nail, with its wooden cupboards and spices and coffee percolator. There is my banjo listing seductively in its kitchen corner. I pick it up and I play: *Jack o Diamonds, Jack o Diamonds, I know you of old. You robbed all my pockets of silver and gold.*

Sweet Sunny South

> Take me home to the place where I first saw the light.
> To the sweet sunny South take me home.

A.

The Eastern Shore of Maryland is not the South of course, unless you lived there. Then you could not tell it from the South. Danny, I have no memory of where you were from. Forty years ago, when we knew each other, I probably never asked you. Now I want to know. Someday I expect we will meet again. When we do, I will ask you where you are from and where you went and what your life has been.

My best friend, Gay, lived down on Quaker Neck Landing Road in a renovated chicken coop with tiny rooms built onto it by her father, a carpenter. In the woods not far from the little whitewashed house stood an outhouse with a pond of ordure oozing out the back. Gay had lived in that little house for as long as I had lived down the road, but perhaps the Browns were not natives of the Eastern Shore but had come from some other place, just as we Longs had come from the Pennsylvania Dutch country of Bucks County, Pennsylvania. The Browns were Catholics, but not Irish Catholics or Italian Catholics. That's all I know.

They were the warmest, kindest people I ever knew. Mrs. Brown, round and soft in her apron and warm brown eyes. Barbara, the big sister, was married and lived down the road. She and her sweetheart husband would come to visit, always holding hands. Little Annie was fat. Gay was thin and pretty with thick curly black hair. Her brother Carl was black-haired and thin and handsome—I had a crush on him. Summer days, Gay and I would walk down Quaker Neck Road through the woods to Quaker Neck Landing and go swimming beside the old wharf. We would saunter back to the house in the late afternoon. Gay's father—lanky and bony—would be sitting in the kitchen, playing his banjo.

Gay died of leukemia when we were eighteen, and I have never been back. Sometimes I think I made up the part about Mr. Brown playing the banjo. But I don't think so.

B.

Music makes a good home when you can't go home. Music makes a good home when your father is too angry to go near and your husband is busy, preoccupied, or else watching the game. Music makes a good religion when the church you grew up in turns out to be more bigoted than faithful and when in any case you are in your raging atheist phase. Music surrounds you and fills you with happiness. Old-time music slakes the thirst of your soul. In those years, Danny, I had a thirsty soul. Odd that old-time fiddle tunes were called the Devil's Music. The devil's music makes a good home and a good family with a long genealogy to go with it. We have Dock Boggs and Fred Cockerham and Tommy Jarrell for grandpas. We have Aunt Molly Jackson and Sarah Ogun Gunning and Ola Belle Reed for grandmas. Hazel Dickens is our sister, Joe Val our brother.

A family reunion takes place ever year at the Brandywine Old Time Music Festival at Brandywine, Pennsylvania, not far from where I was born. Every July at Brandywine, banjo pickers and guitar pickers and fiddlers jam on the meadow, jam in the woods, jam beside their tents, jam in the road.

Jam in the parking lot, beside their cars and pickup trucks. Play "Cripple Creek" and "Sally Ann" and "Sally in the Garden" and "Sally Goodin" and "Sandy River Belle" and "Sugar Hill." Pluck "Cluck Ole Hen" with the fiddle clucking and the banjo plunking. Jam all day under the trees, jam all night under the stars.

The house of music stands and it rings with an old-time tune. To jam is to enter the house of music, to be at one with the universe. To jam is to return to absolute symbiosis, to a prenatal state of communion, to be at one with the other pickers behind this hot picker's pickup truck. Jamming, you play tune after tune, the banjo clucking to the sweet plaintive voice of the fiddle. A tune lasts a long time, and in the pause between tunes there is an exchange of satisfied grunts and small comments as if we had just shared a huge meal, and then on to the next tune. Whenever you want, you retire to your pup tent and crawl into your sleeping bag and let the all-night jammers lull you to sleep like a mama.

You are home and this is an American home. You can march against the war in Vietnam, you can rail against the government, you can petition, you can sit down, you can get arrested. The instrument you hold is an American instrument. Or maybe it was an African instrument (the banjar) made into an American instrument. The fifth string of the banjo drones like a Scottish drone. The thumb string rings its one high note—the drone—among the showers and sprinkles of other notes. The African drum is played, not only by rapping the knuckles on the skin head, but in the very syncopation of the clawhammer manner of strumming an old-timey banjo.

Down in the Valley

If you don't love me, love whom you please.
Throw your arms 'round me, give my heart ease.

A.

There is a painting hanging in the kitchen of the American house of old-time music. The painting shows an older black man sitting on a straight-back chair with a barefoot boy on his knee. The boy is holding the banjo, and the man is chording for him, high on the neck. It's a simple kitchen, a downhome kitchen. There's a tin coffeepot on the floor, a jug. The colors are umbers and burnt siennas, the warm yellows of evening lamplight. The boy, who has golden brown skin, is strumming—his hand distinctly shows he's playing clawhammer style—and his face holds an expression of utter concentration. The musician's blacker-toned face is different. There is concentration—he is studying the boy's strumming hand—but also tenderness. This is *The Banjo Lesson*. Henry Ossawa Tanner painted it in 1893. A reproduction of *The Banjo Lesson* hung on my kitchen wall for all the years I lived in Somerville, Massachusetts. My kitchen always had two banjos, Tanner's old banjo and my own.

B.

In this scene we are sitting on the steep meadow of a mountain in southern Virginia. You, Danny, are not here. I am sitting on a picnic blanket with two of my good-old old-time music friends who have moved from Boston down into Ralph Stanley country. Roads curving through mountains crawling with kudzu, roads twisting through hollers, twisting through coal towns.

We are sitting on our meadow grandstand along with hundreds of others. Whiskey is part of it, beer is part of it; the crowd is growing happier. On the stage set up at the foot of the long slope, bluegrass bands and country bands have been playing since morning.

Now it is dusk, nearly time for the Tom T. Hall band to come on. The sun is dipping behind the mountains. Suddenly the crowd stands up and looks up. We are in dark shadows, and the sky behind the mountains is rosy red. A black speck appears high in the sky. Slowly, but very slowly, it gets larger and larger, mutates into an enormous red-winged angel descending with grace and majesty as if at the Second Coming. It is an extravagantly beautiful sight. The skydiver lands with exquisite grace and the crowd sighs with rapture.

Now it is time for the Tom T. Hall band to come on. The band is got up in red suits. They are setting up on the stage, and people start lighting bonfires on the mountain. This is a downhome Dickenson County festival, no outsiders here, no water truck, no fire regulations.

Bonfires flare on the mountain. The band begins to play. Night falls black and eerie to the slide of a Nashville guitar. Firelight lights the faces. The twang and yodel mix with the hiss of burning logs. The band plays on and on. They sound exactly like themselves on the radio. They must have played these songs a thousand times.

Late late at night the bonfires are still roaring on the mountain. I leave my friends and wander down to the stage. The band is absolutely drunk, barely standing. Yet the songs continue to sound like country music radio. Groupies are there, pretty girls, young pretty girls, and they are going right up onto the stage making dates with the guitar player, the bass player, the banjo.

These boys can play these stale old songs falling-down drunk but something tells me these girls will get a piss-poor night of love. Why do they do this to themselves? Why do girls become groupies?

You Had Some, But You Ain't Gettin No More

> *T'en a eu,*
> *mais T'e n'auras plus.*

A.

Back in Cambridge, Massachusetts, we play one old-time tune after another. Tune after tune after tune. No deviation. We play every Sunday afternoon on Cambridge Common, play like a stuck record. Week follows week. I'm going deaf running a printing press eight hours a day. I'm losing my low notes. Everything gets quieter. I am pissed. I am bored. But I do nothing. I show up at the old-time music jam session on Cambridge Common every Sunday afternoon. I make a pretty good sound, but I'm no musician. Why didn't I put on my walking shoes? Why didn't I take piano lessons, or voice, or learn to play the drum? I did nothing. I was loyal, loyal to my roots, loyal to my friends, loyal to nothing. I was stuck and I stayed stuck. Three years went by. Then I quit.

B.

Thirty years later I tune in to the old-time radio station. The same pickers are picking the same old tunes. I am tuning in to 1973. Nothing has changed. I like it. I pick up my old banjo and I play the only tune that it can play: *Ain't nobody's business but my own.*

I'll Fly Away

> I'm as free a little bird as I can be.

A.

One night remains in mind. On this night, Danny, we Rounders and Rounder hangers-on are staying at the home of some record company just outside Washington, D.C.—Was it County Records? Rebel Records? During the hot July days we sold traditional-music records on the mall at the Smithsonian Folklife Festival. Selling records under our concession tent, we soaked up the music of Africa, Louisiana, Trinidad, Rio, southern Virginia, Harlan County, Lubbock ... At night we returned to this house—Philo Records?—exhausted, but pleased with ourselves. We ate our peanut-butter sandwiches, drank our beer.

On the night I am remembering, you and I took our sleeping bags out through a set of French doors into a small walled garden. It was night and the stars were bright and the grass was dark and soft. We could see the constellations through the branches of a crab apple tree. We unrolled our sleeping bags and crawled in and talked a while and then fell to sleep.

B.

A mockingbird woke us from our dreams. Moonlight flooded the garden. The mockingbird was loud! He whistled, clicked, called, warbled, trilled. His set was lengthy, piercing, shrill. He was raucous as a mean ole fiddle, and devilish, blatantly stealing every other bird's tune with an extravagant flourish. The mockingbird may be a thief, but there's no mistaking a mockingbird. This one-mockingbird string band vibrated so loud and so long that night itself became a mockingbird. Our dreams careened and rollicked along on that crazy old fiddler's tune. I wonder, Danny, if you remember that mockingbird. I will always remember it, and I will always remember your face.

9

The Musician

What is articulated strengthens itself and what is
not articulated tends towards non-being.

CZESLAW MILOSZ

When I was thirty and living in Boston, I spent innumerable evenings
hanging in bars listening to bands—country, bluegrass, Texas Swing,
Appalachian fiddle. At the time—this was 1975—my husband and
I were living separately. Boston was the country music capital of the
North, a cultural anomaly born of the New England port city's situation in the northern foothills of the Appalachian chain. The mountains make a sound-runnel that pours the yodel and twang of Nashville
radio into Boston neighborhoods clear as any local station. One night
I went to hear a certain bluegrass band that, local or not, was good
enough to win notice throughout the South and even in Nashville itself.
The mandolin player was fast fingered and honey tongued, and I took
him home with me that night.

This mandolinist was reed-thin and tall with a wide mouth and a
bush of pale hair and long musician's fingers that scattered notes into
air like magic. He would look down from his height and widen his
mouth in the barest suggestion of a smile. This enigmatic smile, like
that of the Buddha or the Mona Lisa, seemed to contain perfect understanding, perfect knowledge, the subtle perceptions and arcane truths
of a bard. And his mellifluous tenor rendition of some old Louvin
Brothers song—say, "You've already put big old tears in my eyes / Must
you throw dirt in my face?"—seemed to confirm some sort of tender
compassion.

It is likely of course that this good-looking man that I apotheosized was utterly conventional but for his musical gift. Is that an oxymoron? And yes, he was gifted, but was he the genius I thought he was? Was he Beethoven? Was he John Coltrane? I believe that he cogitated extensively on his finances, on his contracts, on the motor in his vw van. It is likely that, contrary to my fervent belief, he had no special insight into life or into love. He was no doubt perfectly ordinary, and why should he not have been? Yet when he sang "When I Stop Dreaming," his intense blue eyes seemed to reflect exquisite knowledge of my own heart.

On the night in question, in the noisy, crowded Harvard Square bar, I felt his eyes upon me. I sat alone on a barstool wearing tight blue jeans and an ironed Oxford button-down shirt with the shirttails hanging out. I had brushed my teeth and brushed my long hair. I felt beautiful. I could imagine us kissing, or sharing arcane secrets known only to ourselves. I could imagine us coming to complex understandings. In our two- or three-week acquaintance before I dragged him home with me, I had told him I was writing a book. To my mind, this was an important attribute. This made me special and unique. I told him I was writing a book several times, for the musician could not seem to remember this important fact, or perhaps he could not imagine in me a person who could write a book. The truth is that this book-writing project was not going particularly well.

We spent the night in my spare, wood-floored apartment with its clean-made mattress on the floor, its novel—*Mrs. Dalloway*—lying face down beside the mattress, its low window giving onto a second-story picket-fenced porch overhung with budding cherry branches. We spent the night passionately, more or less, but strangely, as any such night spent with a total stranger must be. Let's say we spent the night enfolded in my delusions, which were alcohol-pickled just about to perfection. In the morning he was gone. He left with his own life, his own self, his own delusions, whatever they may have been, still in code, undeciphered. He left for good.

I fell into inconsolable grief. For he spoke my own language, or so I felt. He played the music of myself, and in that interlude I had felt visible and recognized. I had come into my own existence, into my own life. Without him, my thoughts, my very being remained without existence in the world, unheard, even by my own self. Why I felt this way, when he could not remember one simple fact that I had troubled to tell him repeatedly, struck even the person I was at the time as slightly off base. In my misery, it dawned on me that most people sip coffee, not beer, with their eggs in the morning. My life had been spinning off course for quite a while, and it finally crashed into a wall. He was the wall. I quit drinking and spent most of that summer lying on the living room floor. Gradually, in this way, I came to an essential insight about life and art.

How it is that the expressive ones—artists, musicians, writers—may carry existence, affect, meaning for those without voice or venue. How these expressive ones may speak for the silent, for those who have not found voice, for those whose voice has been scorned or scolded or even slapped out of them. How badly I wanted to sing my own song is told in how badly I fell for that musician.

"There is always a difference of course," Seamus Heaney says, "between the poetry you would like to write and the poetry you are given to write." Gradually, I learned to write what I was given to write, to put words down on the page, to hear the music of my own voice, to become a poet, to become a writer, to become an author. As for the mandolinist, he kept on making a name for himself in the world of bluegrass music, and if his career ended up as a rather minor blip on the cultural screen, as it seems to me, that is only in comparison to the Christlike figure I had made him out to be.

I picture the musician now—like me he will be past sixty—playing his banjo at some kitchen table in Tennessee, where he moved and has lived ever since. Perhaps it is night and the lights are off and his wife has gone to bed, and he is sending sprinkles and sparkles of notes up to the moon, doing in solitude, for this moment, what he has been given to do.

From that experience, I learned something very simple: No one else carries what is in your own heart. No one else can play the music you are given to play. What I imagined onto that musician was my own potential virtuosity, my own voice, my power—for power it is—to articulate the world, to color it with your own sensibility, to move it and shake it for a short sweet time with your own kind of song.

10

Dressing

Not easy for us
to equal goddesses
in lovely form
SAPPHO

I once had a friend who could wear anything—bikini, business suit, jumpsuit. She was a neat-boned *Vogue* model who dropped into our blue-jean world of the late 1960s as if from a different movie. For a brief time she became one of us, living in the margins, smoking dope, dancing the nights away. She was photogenic, that ephemeral quality that differs somehow from good looks: photographers drooled after her like puppy dogs.

She was no writer. But she created characters out of outfits, going on whim from jet set to waif, from shopgirl to vamp. She was like that Wizard of Oz duchess who spent her days in a mirrored room trying on various heads from her large, fashionable collection.

Our *Vogue* model, too, collected selves. For her, crossing the boundary from one character to another was as simple as dressing for dinner. She also designed clothes, and stitched them. Quite casually, in one afternoon, she made me the most beautiful dress of my life, a short black raglan-sleeved dress with a mock-turtleneck that could be worn to the Laundromat or to any more formal occasion. She imagined for me a lovelier persona than I imagined for myself, and provided the means of transformation.

I'm terrible at dressing, but gradually I've come to find it entertaining. I was brought up to the care of cows and dogs, and as a child I entertained myself by reading the old books that lined the walls of our crumbling farmhouse. We dressed in jeans and in Sears Roebuck rubber farm boots except for Sunday school, a stiff affair. We wore the required dresses to grade school of course—beautiful dresses stitched by my Pennsylvania Dutch grandmother—but unfortunately in the first wash our rusty water pipes stained them with great turd-colored blotches, and our school dresses all looked irreparably soiled.

Our mother was oblivious to dress, a blindness she shared with our father. My twin sister, Pamela, and I considered ourselves vastly superior in matters of fashion. We had regulations and stipulations: We refused to dress alike and looked down on twins who walked around like duplicates. However, we lacked teachers. Our family lived to a large extent outside the cash economy, and shopping we knew nothing about.

Undoubtedly though, our interest in dressing was the legacy of our Grandmother Henry, the mother of our mother. She sewed all her life and not only stitched for Pammy and me all our school dresses but also our flower-girl gowns for our wedding duty. Our favorite school outfit was a calico print dress, daisy-on-black, smocked, with a sash and a round collar with a center-tab edged in yellow piping. I remember my grandmother's little white sewing room and the whir of her Singer sewing machine. I remember her pins stuck in a pincushion, needles in their needle-paper, pinking shears. I remember her gray-felt dress-form draped in its measuring tape. I remember the crinkle of pattern paper and the white pattern envelope printed with color pictures of the dress or skirt or suit.

Despite this inheritance, I'm no seamstress. During my young wife-hood, I sewed for my husband a man's wine-colored India-style shirt with a slit neck and a wine-colored satin lining. He appreciated it deeply, but it was too much for him and he never wore it. I also made sweaters using an instruction book that showed you how to measure a person to make a sweater to fit an actual torso, an actual arm. I knit my mother a navy blue wool cardigan, and she wore it for many years. I knit my beloved grandmother a purple shawl. I asked my father if

I could measure him so I could knit him a sweater, but he declined, explaining politely that he already had a sweater.

So much for my attempts to dress others. Dressing myself has been frankly more difficult because I do not have the stick figure to fit into clothes designed for anorexic models. Besides, I was never inducted into the arcane feminine guild in which elaborate rituals of dressing, I am convinced, are passed along. To this day I have no idea how to tie a scarf.

As for the characters who people my fiction, they dress, for the most part, quite well. I select their wardrobes with attention and care. True, I am burdened with a few curmudgeons who disdain the very idea of good clothes—two are sinewy farmers who have met neither each other nor my father—but I also have characters who dress with a certain impeccable elegance. Maxwell Stern makes a very good living stealing fur coats. When he steps out in his white double-breasted suit and white Panama hat, he could easily make a *People Magazine* Best-Dressed-Man-of-the-Year Award. At one time Edith Steiner wore conventional cap-sleeved sheaths accessorized with pearls, but then her husband dumped her and she in turn dumped her wardrobe along with the rest of her false life, and went to live in the woods. As a hermit, she wore corduroy pants and boots and rustic shirts. She got her garments by carving walking sticks and trading them for duds at the country store.

Dressing characters is a complex task requiring shopping. I have shopped exclusive shops without spending a dime, and once at Neiman Marcus tried on a Russian sable overcoat that went for a hundred thousand dollars. In the same work in which Maxwell steals such an overcoat, Olivia Sharkey plays jazz violin in a three-hundred-dollar T-shirt, a black-gauzy thing scattered with blood-colored velvet leaves, its V-neck lined with black marabou feathers. Maxine completes their triangle wearing blue jeans, a black turtleneck, and white New Balance running shoes, which, unfortunately, call attention to her size-12 feet.

Dressing characters provides entertaining moments, but there are no expensive moments. Dressing oneself is more complicated. As I approached my fortieth birthday, the question of obtaining a decent wardrobe became acute. Basic attractiveness, previously taken for granted, began to seem at once essential and slightly ephemeral. This was during the era when I put bread on the table by operating a printing press. Looking back on it, I am startled to realize that I chalked up a decade as a printer in part because I could print in what felt like my own clothes. In any case, my turn into low-middle age marked the beginning of my emancipation from blue jeans, work boots, T-shirts, and my trusty red-and-black-plaid lumber jacket.

At the start of my campaign for a decent wardrobe, I determined to gather to the cause a coterie of assistants. My first and primary fashion consultant has always been my sister Pamela. Pamela grew up to get a PhD and to become a brilliant historian of technology and culture, with grants and awards falling upon her like rain. But her education in matters of fashion is frankly no better than mine. Nevertheless, she loves to advise me and I love to advise her.

Next I turned to one of my dear friends—this was during the Boston years—who was extremely experienced in all matters of fashion due to her longstanding, extensive, and perfectly trouble-free avocation of shoplifting. One day Isabelle—as I shall call her—brought down her new overcoat. Now she is a pretty woman but was beginning to soften into the little round grandmotherly look common among middle-aged women from her country of origin—let's call that France. It should be further mentioned that Isabelle raised her child in poverty as a single mother. For her bright and funny boy, she was the best, most devoted, most loving mother I have in my life witnessed. But she wanted a man. She wanted a good man. She wanted a lover and she wanted a husband and she wanted a good stepfather for her child. When she put on the cashmere coat for which she had abandoned her shoddy, stained, worn-out, woolen coat—I was shocked. The coat transformed her into a beautiful woman. Shall we look down on her? Shall we cluck our tongue? And—she got the man of her dreams and she is happy, and I am happy for her happiness. I have always loved Isabelle, she is one of the dear friends of my life, but the problem with her fashion advice is that she has no price range. Besides, I am a coward and could not possibly rip off clothes even if I wanted to.

Years have passed, and I am still learning to dress, even if slowly, and with mistakes. Here characters, and not only my own, have schooled me in a subject that some daughters learn from their mothers. I have meditated on Mrs. Dalloway's shopping day. I have considered the manner in which Cinderella's fashion crisis was solved with supernatural assistance. I have admired the contemporary film version of Cinderella— *Pretty Woman*—in which it is a man, the hotel concierge, who plays the fairy godmother. I have noted the fastidious care with which Raymond Chandler's detective Philip Marlowe dresses for work: "I was wearing my powder-blue suit, with dark blue shirt, tie and display handkerchief, black brogues, black wool socks with dark blue clocks on them. I was neat, clean, shaved and sober, and I didn't care who knew it. I was everything the well-dressed private detective ought to be. I was calling on four million dollars."

Most characters dress in order to pass from one realm to another, to cross a border, to escape into a world to which they do not belong. Consider the expression from rags to riches (not from shacks to riches, not from bad teeth to riches). Consider the expression "to dress down," meaning to scold or to reprimand (to expose, to undress, to return, perhaps, to a state of naked truth).

The opposite of dressing down may be to dress up, to dress a part. And why not dress a part? Who is to say that persons cannot become their portrayals? Perhaps the person most in need of faking out is that old shoe—the self. Perhaps everyone stepping out every day is engaged in some form of passing. The big question about the extreme forms— from black to white, say, or from man to woman—is only this: What happens to the old self that has been discarded like a pair of bell-bottom jeans?

Those bell-bottoms—narrow-thighed and slung low on the hip— were to 1966 what the song "California Dreamin'" was to that same era. Indeed, any garment—sock, cap, shirt, or shoe—comes loaded with connotation, with history. Outfits fit both body type and decade. Are you wearing a bra? You belong to the twentieth century, for in the nineteenth it was the corset. A skirt? You could be a man in twelfth-century Scotland. Or not. It is possible to consider the march of centuries and cultures in terms of the hat, from warrior's headdress to top hat to French beret, this last worn by men reading Beckett through wire-rimmed glasses.

Clothing may cover the body, but how much more does it clothe who we are and who we wish to be. If all the world's a stage, our garments are the costumes and the props. Which of my selves shall I wear today? Which face shall I put on, to meet the faces that I meet? What does my dark side wear? My child? How do I dress my wounds? Perhaps we each have an original self, the one we go home to. Mine is a farm girl dressed in blue jeans and work boots and a plaid flannel shirt. I once had a friend who worked as a nurse, but on her Saturday off, she went around as Louisa May Alcott's Old-Fashioned Girl. She wore an ankle-length cotton-print dress with long sleeves and covered buttons. How different is depression's outfit—dull-colored and ill-fitting and unkempt. How different again seduction—draped in a slinky red dress. It is possible to go out dressed in a self that is entirely unused to company—a Queen, say, or a vamp. Then we may have the sudden urge to escape out the back door of the party and run home. Our grandiosity has been exposed. We were posing as royalty when the truth is we are the cleaning lady.

For an outfit also has an inner fit. It reflects, whether unconsciously or willfully, the inner mind, the inner spirit. Georgia O'Keeffe, who designed and sewed all her own clothes, wore simple dresses and long skirts, always black. She costumed a persona: serious, mysterious, an artist. And who is to say that a consciously created persona is not as valid as any other? Coco Chanel began as an awkward, homely girl. But she learned to dress in elaborate costume, as if entitled, and became the empress of fashion. Was she really homely then? Or was she really beautiful? Or was she homely until she made herself into a beauty? She had courage, and indeed courage is required to dress, to dress up, to cross-dress, to dress down. To dress is to impose one's own definition of one's own self upon the social scene.

To construct a persona—a public personality in Jung's terms—is in some sense to construct a person, a personality. This is exactly why gifts of clothing are so loaded: someone else gets to define who they think you are—or should be. That my grandmother made us school dresses that we loved proved to us that she loved us.

The persona I bring out when I read my poems to an audience is as follows. I wear clean blue jeans and some sort of black-turtleneck top. In other words I stand behind the poems in a somewhat spiffed-up version of my original self. I also wear my amulets—in one pocket a stone of coal, around my neck a Yin Yang circle carved in cowbone, on my ring finger a blue stone set in silver wire, not to mention my secret good-luck underwear.

But that outfit didn't fit the occasion of a reading at Seattle's Elliott Bay Book Company for the contributors to *The Feminist Memoir Project* anthology. I wanted my feminist to look feminine, understated, somewhat expensive. She wore a narrow black skirt, a black T-shirt, and over that a raw-silk coppery-gold jacket with raised shoulders and patch pockets. This last had consumed what was for me at the time the staggering sum of $285. However, I consider her costume to be my best sartorial achievement so far. The reading turned out to be a well-executed occasion full of lively talk. And—a friend in the audience later told me: You looked rich.

Clothes advertise class, region, occupation, conspicuous wealth, conspicuous absence of wealth. In fact, it is difficult to imagine any material or immaterial situation that is not in some way connected with clothes. We fight in uniform, bask in bathing suits, sleep in pajamas, run in sweatpants. Even death insists upon its grave clothes. Our clothes connect us by the very skin to the natural world. This was brought home to me when my friend Maury Klein presented me with the gift of a red-fox stole—we were rummaging in the attic of his big old Narragansett, Rhode Island, house. He had inherited the stole, which retained its fox head, fox tail, fox ears, fox paws, and small fox face, from one of his father's inebriated ex-wives, an ex-fashionable woman of the 1940s. I declined the gift. By then I had moved to the other side of the country, and I explained to Maury the danger of walking down a Seattle street—we are very pro-animal here—with a dead fox warming my neck.

Still, I wear shoes made of cowskin, carry a shoulder bag fashioned from a dead lamb. Until 1924, when DuPont invented the artificial silk, nylon, every fabric on every back was taken from a plant or an animal: silk spun by moths, wool sheared from sheep, mohair and cashmere clipped from goats, cotton plucked from the cotton plant, linen spun from flax.

Dressing is complicated by fabric, cut, drape, and fit, and I'm a slow learner. At Yazdi, a store in Seattle's Wallingford Center, I spend an hour deep in rayon scarves—gold-flecked purple or black or peach satin, this last an evening gown with spaghetti straps that I would not myself be caught dead in. Yazdi sells slinky dresses, bulky sweaters, ropes of glittering beads. Next door, another shop purveys jumpers woven from hemp, and cotton jackets and skirts, sturdy and lovely looking. I support ecological, plant-based clothing and once or twice a month go there to try something on, only to find in the mirror a rather dowdy-looking frump scowling back at me.

Frump or not, I have found a wardrobe mentor. Her mother, too, was a fashion dud. She has been where I have been, but she took a different path out. She is a former model, a personal shopper, a therapist whose erudition in matters of dress is acute. She is no clotheshorse, but rather a professional who thinks about how this person feels in these clothes. She studies the fit of garment to person. Barbara Blackburn is writing a book (I coach book writers) on beauty, on the possibility of anyone owning the beauty that is hers to possess.

I also found the perfect store. Opus 204 was an elegant but comfortable place in downtown Seattle, with scarves and shoulder bags arranged on oak tables, with crockery for sale along with somewhat expensive (they had good sales) dresses, skirts, coats, trousers, vests, and jewelry. The main saleswoman was a cheery, posh, rather heavy woman who wore my name—Priscilla—in rhinestones above her right breast. Once when I was trying on some blousy thing she emitted a scream from the other side of the store—No!—and came running over with something different to try on. I immediately fell under her spell. At Opus 204 they knew the arcane secret of making rather sturdy-looking figures such as mine look slender: shoulder pads, and slightly longer,

straight-cut blazers that button all the way up. The store had a work-
room in the back where dressmakers designed and sewed most of the
garments they sold.

Opus 204 was my doorway into "good clothes." At Opus I bought my
first truly worthy garment. I thought I was buying a tunic-length blazer,
but after wearing it a few times I realized that it was not a blazer but a
shirt. (I am fashion-blind the way some people are plant-blind, unable
to tell shepherd's purse from purslane although they look at these
common weeds every day.) I tried on this shirt, made of herringbone-
woven wool, shoulder-padded and buttoned up to a simple collar. It
was on sale for seventy dollars, reduced from three hundred. There was
something about the cut. In the mirror I saw understated good looks.
I saw a rather attractive woman who was perhaps competent at what
she did, no doubt rather well paid. I actually looked behind me. No, this
was me. I bought the shirt. It was not too long afterward that I began
asking for raises at my copyediting job, and began receiving them. That
shirt, I am convinced, paid for itself.

I'm nowhere near death, I hope. But I like to wander into thrift shops,
if only to think of all the characters who have shed their vestments,
their *in*vestments, whether out of whimsy or some larger transformation. Here is a red satin slip and matching red pumps. Here is a
black sequined gown. What glorious evenings, what romances have
these castaway garments seen! What buttons have been unbuttoned
by whose fingers, what clasps unclasped! Whatever lives and whatever
loves these clothes have passed through, all was ephemeral, in the end.
Whatever glad-hearted girl first wore this olive-green floor-length
gown cut in the flapper-style of the 1920s has no more need of clothes.

She has gone from this world, disappeared entirely from the world
of appearances. And I, in high middle age, have discovered something
new about appearances that I would not have guessed even a few years
ago. How it is that a person may disappear *into* the world, before dis-
appearing *from* the world. How the world, early one morning, becomes
all rain and whispering leaves and the padding and purring of cats.
Without the self in it to bother with. Just the world itself, for whole
moments of time.

And how, more and more, you drop the need to "arrange your face to meet the faces that you meet," as T. S. Eliot wrote when he was only twenty-five and didn't know any better. Especially no need to arrange your face to meet the faces of your friends. For who are your friends if not the ones who love you for who you are?

And what of that friend of long ago, the *Vogue* model we began with? I imagine her on some white beach, turning this way and that for the camera. Perhaps white cliffs rise up behind a turquoise sea. She changes from bikini to beach robe back to bikini. This is her job and she is good at it. Now she is done for the day. Her image, once again, will flit through a hundred glossy pages, unrecognized. I have looked for her face and it has become lost among the faces. I hope she has become very rich and I hope she is happy in love. We can imagine her stepping out of one guise or another into blue jeans and sandals, into her original self, the one we knew for a short while so long ago. She of the frenetic social life has been looking forward to this evening. She goes back to her hotel, dines in her room, pulls out of her suitcase a thick novel. She settles down for the night to read. We will leave her there, lost in the world of the imagination, that near, familiar world where the characters—so like ourselves!—all wear their hearts on their sleeve.

11

Genome Tome

Suddenly all my ancestors are behind me. Be still, they say. Watch and listen.
You are the result of the love of thousands.

LINDA HOGAN

Our ancestors are behind us. They gave us our lives. We carry their
genes. Whoever we are includes part of whoever they were.

But who were they? We know and we don't know, but we know a lot
more than we did even a few years ago. The scientific revolution known
as the Human Genome Project began in 1990 as an international effort
to map the human genome. With jubilation scientists announced in
June 2000 that they had completed a rough draft. By 2003 they had discovered most of the estimated twenty to twenty-five thousand human
genes found on our double-strand of twenty-three chromosomes. This
chapter is a montage with twenty-three sections, one for each pair of
chromosomes. It was inspired by a 2002 art exhibition titled *Gene(sis):
Contemporary Art Explores the Human Genome* mounted here in
Seattle at the University of Washington's Henry Art Gallery. The exhibition sent our town into a flurry of lectures mutating into poetry read-
ings mutating into PowerPoint presentations of elementary genomics.
My obsession with the subject began then, and it has continued to the
present day. Its deep origin has to do with my own genome. I am an
identical twin—one of nature's clones.

Grandmother

Four to eight million years before we were born (before any of us were born) there lived in Africa a great ape, which our species has termed "Pan prior." Out of "Pan prior" (in quotes, since bones have neither been found nor designated) both chimpanzees and our own line evolved. This process of evolutionary change occurred in a manner scientists term "messy"—lengthy and complex. Over millions of years various apelike species evolved, and among these populations several cross-hybridization events occurred before our two lines—chimp and human—separated for good. Still, our primal primate ancestor was a chimpanzee-like ape.

This grandmother ape, how shall we think of her? Shall we despise her as if she were a massive piece of crud in our shiny kitchen? Shall we deny that we have inherited her genes? Shall we strut about as if we ourselves were made of computer wire and light?

Corps of Discovery

The Human Genome Project is the Lewis and Clark Expedition of the twenty-first century. In 1804 Meriwether Lewis and William Clark and thirty-one other souls (the Corps of Volunteers for Northwest Discovery) traveled into a country that was to them entirely unknown. They traversed rivers, mountains, prairies, swamps, rapids, cataracts. They took specimens and made notes and drew maps. Mapping the human genome—from twenty to twenty-five thousand genes strung along twenty-three pairs of chromosomes—is also to journey into the unknown. Lewis and Clark meant to befriend the Indians, but in the end, they cleared the way for the destruction of indigenous ways of life thousands of years old.

As human genomes are mapped, as the genomes of mice and flowers and fleas are recorded, much will be revealed—the secrets of life itself. And much good will come. Cures. Vaccines. Forensic evidence. But make no mistake: There will also be difficulties. There will be dangers. What if an insurance company discovered which of its customers carried the gene for that very expensive disease, Parkinson's? What if some authority drew up a list of all persons, including the innocent, who carried the two genes that make violent acts more likely? What if we were to learn of certain untoward anomalies in our parents' sex lives, anomalies that alter forever who we think we are? (True story: A daughter took a cheek swab from her dying father in order to get the DNA for her paternal line. After his death the results came back. This was not her biological father. Was there anyone still alive who could explain this shocking result? No.)

Alba

Take the gene that produces fluorescence in the Northwest jellyfish. Inject the green gene into the fertilized egg of an albino rabbit. Get Alba. Alba, the green-glowing bunny. Alba, designed by an artist in Chicago, created by a lab in France. Alba, a work of art, a work of science. Alba, the white bunny with one strange gene. Alba's jellyfish gene made Alba glow green under ultraviolet light. Oh Alba. Oh funny bunny. Oh unique creature. Oh sentient being without fellow being. Oh freak without circus, star without sky, noise without sound. Was Alba a work of art or was she just one more lab animal? After a lengthy dispute between Alba's artist and her scientist, her scientist announced her death.

In Greek mythology a chimera was a fire-breathing animal with the head of a lion, the body of a goat, and the tail of a serpent. In our time a cross-species chimera is an animal created from the cells of two different species. Today many labs create chimeras for purposes of basic research or to find a cure for cancer or to develop replacement organs or cells. This research is regulated. Chimeras are not let out the door or permitted to reproduce or even, often, to grow beyond a few cells. Still, the possibilities are there. "Man or Mouse?" asks a document produced by the Danish Council of Ethics. "Will chimera research be capable of producing crossbreeds that cannot be classified as either animals or humans?" What of animals whose cognitive abilities have been altered in a human direction?

Recombinant Recipe: Spider Silk

The spider web is the strongest natural fiber in existence. But for centuries attempts to raise spiders in the manner of raising silkworms have failed, due to the spiderly taste for other spiders. Spiders eat spiders eating spiders.

The genomic solution: introduce a spider gene into the goat genome. Spider-goats in their spider-goat barns look like goats—curious eyes, heads cocked to one side, perky ears. Their milk, strained like cheese, spun like silk, could in theory produce a filmy fabric, lightweight, stronger than steel, softer than silk. But milk-silk is a work in progress. Transgenic goats are being bred, but it's not easy to produce commercial fabric from spider-milk fiber. No one has yet done so. Another idea: transgenic silkworms. Put the spider's dragline gene into silkworms that already spin silk. Someday spider-silkworms might spin a fiber that could be woven into a bulletproof vest.

Or a beautiful, bulletproof, spider-silk dress. If this happens, I want one.

Next of Kin

Chimps have long arms for climbing and for swinging in trees, and they have opposable thumbs and opposable big toes. They knuckle-walk—walk on all fours with their hands folded into fists. They are born with pale faces that gradually turn brown or black.

Chimps live in large sociable communities that have an alpha male and several (less dominant) alpha females. They express affection by grooming each other with obvious pleasure and elaborate precision (they can remove a speck from the eye or a splinter from a toe). They can be quite aggressive—communities have been known to go to war. Chimps are territorial, and when they happen upon an isolated foreign individual on their border, they kill. Like humans, they are capable of cannibalism, of infanticide. But chimps also laugh and kiss and hug. They dine on a diet that varies from plants to ants, using stick-utensils (tools) to work the ants out of the ant-cupboard. During the day they spread out in small groups to forage for food. While they are thus scattered the males drum, stamp, and hoot—the chimpanzee Global Positioning System. At night they gather and make nests high in the trees. When a chimp is born, the other chimps come around offering to groom the mother for a chance to inspect her baby. Mother chimps are fiercely attached to their infants. Baby chimps suckle for three to five years. Adolescents stick with the family and help to babysit the little squirt. The baby requires a long time—five to seven years—to learn all the ways of chimpanzees from chimp-talk (so to speak) to tickling to hunting food to building the nightly nest. A chimpanzee becomes an adult between eleven and thirteen years of age, and can live to the age of sixty.

In December 2003 a chimpanzee genome was read for the first time. Chimps, genetically speaking, are remarkably similar to humans. They are practically family. Indeed, scientific reclassification based on such revelations puts them *in* our family. In 2016 the term "hominid" refers to the family (Hominidae) of all great apes and their ancestors: humans, chimpanzees, gorillas, and orangutans. The term "hominin" now refers to the genus that includes all species of humans and our immediate ancestors (including members of the genera *Homo, Australopithecus, Paranthropus,* and *Ardipithecus*).

Lament for Ham and Enos

In the late 1950s the United States Air Force acquired sixty-five juvenile chimpanzees. Among them were Ham and Enos. No doubt Ham and Enos and the others had witnessed the slaughter of their mothers.

Let the new life begin. The Air Force used the chimps to gauge the effects of space travel on humans. The small chimps were spun in giant centrifuges. They were placed in decompression chambers to see how long it took them to lose consciousness. They were exposed to powerful G forces—forces due to acceleration felt by pilots or by riders on roller coasters.

Three-year-old Ham was the first chimpanzee to be rocketed into space. This occurred on January 31, 1961. NASA archives record "a series of harrowing mischances," but Ham returned alive. The results pleased astronauts and capsule engineers, and three months later Alan Shepard became the first American to be shot into space. Ham was put on display at the National Zoo in Washington, D.C., where he lived alone from 1963 until 1980. He was then moved to the North Carolina Zoological Park in Asheboro, where he lived with other chimps. Ham died on January 17, 1983.

Enos, age five, was launched on November 29, 1961. Enos had undergone a meticulous year of training to perform certain operations upon receiving certain prompts. But upon launch, the capsule malfunctioned and Enos received an electric shock each time he acted correctly. Nevertheless, he continued to make the moves he knew to be right, shock after shock after shock. He orbited earth two times and returned alive. Eleven months later, Enos died of dysentery.

The following year John Glenn orbited earth three times. On March 1, 1962, in lower Manhattan, four million people greeted Glen and two fellow astronauts with a huge ticker tape parade, confetti falling like snow.

The other chimps, the chimps captured with Ham and Enos, were transferred to "hazardous environments" duty. To test the new technology of seatbelts, they were strapped into sleds, whizzed along at thirty, fifty, one hundred miles per hour, slammed into walls.

By the 1970s the Air Force, done with the other chimps, leased some of them out for biomedical research. These highly sociable primates, now adults in their twenties, were stored in cement-block cells with bars in front, but with no windows between cells to provide contact with fellow-chimps. The Air Force retained others in cages. A few were rescued and brought to chimpanzee sanctuaries.

Chimpanzees are our nearest relatives in the primate world. And they do suffer. There are some differences though. After his flight, Ham was photographed "grinning." But in the chimpanzee face, this expression is a fear grimace. Of Ham's "grin," Jane Goodall said that it showed "the most extreme fear that I've seen on any chimpanzee."

Lucy in the Sky with Diamonds

The fossilized skeleton of Lucy, discovered in Hadar, Ethiopia, in 1974, was the oldest hominin remains then known. Lucy died 3.2 million years ago. As her discoverers, Don Johanson and his team, were that evening looking at her bones in amazement, a Beatles tape played in the background. They named Lucy after the Beatles song "Lucy in the Sky with Diamonds." Lucy was short, about four feet high, with long arms for climbing. She stood upright. That's the important thing. Her proper species name is *Australopithecus afarensis*. From her group, several species of hominims evolved. *Homo erectus* evolved. We evolved. That's the old story. It's a nice story. It has a nice beginning, middle, and end.

But paleoarchaeology gets complicated. Bones speak, but they do not enunciate. Skulls and femurs and molars are measured and compared and recompared, and theories replace theories. Thighbones and skulls "from the same species" placed side by side look terribly different, and fossilized bones, alas, do not often produce DNA.

About six million years ago, our human lineage parted from the chimp lineage (a fact told by molecular data). So, about then, the first hominin—meaning human or direct ancestor to human—must have lived. This first hominin has been the longed-for find, the physical anthropologist's Holy Grail. In Kenya, in 2000, scientists Martin Pickford and Brigette Senut discovered a very few very old bones. *Orrorin tugenensis* (Orrorin means "original man") lived six million years ago. This being was likely bipedal, which is diagnostic for human. (And no, it did not live in a savannah. It lived in the woods.) It could also climb, and had other more chimplike features.

But wait. There's a second candidate for this first hominin— *Sahelanthropus tchadensis*, nicknamed Toumai ("Hope of life"). Michel Brunet and team discovered Toumai in Chad in 2001. This being lived between six and seven million years ago and had small canine teeth, also diagnostic for *Homo*—human. Male chimps and other primates have long canine teeth that they use to threaten and harm other males. Our kind just does not bare its teeth to express aggression. Toumai was likely bipedal, although this is debated.

Then we get to the *Ardipithecus* genus (*Ardipithecus kadaba* lived 5.7 to 5.2 million years ago, and *Ardipithecus ramidus* lived 4.5 to 4.2 million years ago). They are more definitely our ancestors based on bone evidence for bipedalism and also the configurations of jaw and teeth. These lead to the several species of *Australopithecus* including *Australopithecus afarensis*, that is, Lucy. It is pretty well agreed that our genus, *Homo*, evolved out of an *Australopithecus* species between 3 million and 2.5 million years ago. But this leap from *Australopithecus* to *Homo* is shrouded in lack of evidence, lack of bones, lack of knowledge.

Consider *Kenyanthropus platyops*, "flat-faced man from Kenya." This man lived 3.3 to 3.5 million years ago, according to the extensive website on human origins published by the Institute of Human Origins, founded in 1981 by Don Johanson of Lucy fame. Okay. There may well have been several species of hominins running around Africa. We are not descended from all of them.

Meanwhile, scientists have unearthed three different species of the genus *Paranthropus*, who lived 2.7 to 2.5 million years ago. Unresolved phylogenetically. Definitely related to *Australopithecus*, of which there were several.

Now we get to *Homo*, our genus. The old story goes like this: *Homo habilis* ("handyman") to *Homo rudolfensis* (both lived just over two million years ago, and both had larger brain cases). Then we have *Homo erectus* out of which evolved *Homo heidelbergensis* out of which evolved, in Europe, *Homo neanderthalis* and much later, in Africa (two hundred thousand years ago) *Homo sapiens*—our people.

A good story, with a nice narrative arc. But is it a true story? Scientists are beginning to question the very existence of *Homo heidelbergensis* as a distinct species. And it may be that *Homo habilis* (represented by very few fossils) was actually an *Australopithecus*. *Homo rudolfensis* may also have been an *Australopithecus*, this one with a larger brain. And now in South Africa, Lee Berger and team have uncovered a new species of human, represented by hundreds of bones, in the Rising Star cave, not far from Johannesburg. *Homo naledi*. It is unquestionably *Homo* but has features of *Australopithecus* (it is as yet undated). Might *Homo naledi* be the original *Homo* species? It might be. Though Donald Johanson of Lucy fame thinks these bones may be *Homo erectus*.

I like what Jeffrey H. Schwartz and Ian Tattersall have to say about this "morass of specimens" in an August 2015 issue of *Science*: "if we want to be objective, we shall almost certainly have to scrap the iconic list of names in which hominin fossil specimens have historically been trapped, and start from the beginning."

Where do we come from? We don't know, not yet. But we're getting closer . . .

Eve, and Adam

All seven billion of us *Homo sapiens* have a common mother; let's call her Eve. We know of the existence of this unique, single matrilineal ancestor due to the DNA in mitochondria. Mitochondrial DNA differs from that in our genome proper, which lives in the nucleus of the cell. Mitochondria are small organelles that exist outside the nucleus, in the cell's cytoplasm. The mother, not the father, passes mitochondria to both sons and daughters. Mitochondrial DNA mutates at a steady rate over the generations and in this way becomes a time machine back to the original mother of us all. She was not the only woman in the world, and neither was she the first woman in the world, but she is a common matrilineal ancestor for every single human being alive today. All other matrilineal lines have died out.

And what of Adam? The Y chromosome exists only in males and is passed from father to son. This chromosome contains a long stretch of non-combining DNA that is passed down through generations of males, passed down exactly as is, except when it mutates. This, then, serves as a second molecular clock. We also have a common father; let's call him Adam. He was not the only man in the world and neither was he the first man in the world. Y-chromosome Adam and mitochondrial Eve did not necessarily live at the same time. They never even met, much less got it on. It used to be thought that our Eve lived much earlier than our Adam. Now we think they lived within the same few thousand years.

This is the news revealed by the book of the human genome, the book whose pages we are still learning to turn.

Genome Tome   103

History and Geography

We are apes evolved into humans. We are Africans, genus *Homo*, species *Homo sapiens*. We evolved in Africa from some earlier *Homo* species about two hundred thousand years ago. Not so very long ago. Ten or twelve thousand generations ago.

We are *Homo sapiens*, alone knowing. We know and we don't know. We wonder. We wonder where we came from. We wonder who we are. We wonder where we are going. We pose questions.

Questions

Are we, then, the greatest of the great apes?

Is human kindness more human than inhuman cruelty?

What makes a cell divide? Am I dividing against myself?

If we were once single-celled creatures, was I once a single-celled creature?

Identical twins—aren't we the pioneer clones?

How does Earth's age—4.6 billion years—relate to our age?

If grammar is innate, is iambic pentameter innate?

If you could read the book of your genes, would anything there surprise you?

Would it surprise you to learn that you were mixed race?

Can humans and chimpanzees mate?

What will life look like after five hundred years of genetic experiments?

Is human selection less natural than natural selection?

Where did we come from? What are we? Where are we going?

If a twin is not the same person, why would a clone be the same person?

Should art include the creation of life?

Is there a gene for creativity, and if so what protein does it express?

If a scientist creates a new species, is the scientist the parent? Who gets custody?

Do I belong to myself, in the cellular sense?

Who wrote the book of life?

Is my cell line mine? Is my genome mine?

Considering that more genetic variation exists within racial groups
than between racial groups, what is race?

Was our first mother happy?

You said—what?

The Grammar Gene

Linguist Noam Chomsky argues that grammar is not learned but somehow comes with our DNA. People in any language recognize grammat-
ical structures, apart from the sounds or meanings of words. Grammar
is innate, whereas diction and meanings are cultural and, over the slow
centuries, in flux. Others argue that what is inherited isn't grammar, it's
a propensity to search for patterns in speech. We move from Mama!
to Mama get ball! to I think Johnny went to the store to get milk, at
least that's what he said he was going to do before he found out he won
the lottery—a construction that will forever elude the most brilliant
chimps taught to "speak."

Did language evolve out of primate vocalizations? Or did it evolve
out of an entirely different part of the brain, the part that can practice
throwing to improve one's aim, the part that can plan to marry off one's
unborn daughter to the as yet unconceived son of the future king.

There is something about language that we inherit. Perhaps our
mother taught us to speak, but she could never teach a chimp to speak,
except in the most rudimentary way after years of work. We are born
with something structural about language in our DNA.

The structure of language lurks below the meaning of words.
Chomsky wrote, "Colorless green ideas sleep furiously." This grammati-
cal sentence illustrates that grammar and meaning have about as much
relationship to one another as two strangers on a blind date. Grammar
is the townie. This dude, this thug knows the ins and outs of the place
by heart. He runs the show, and he practically owns the territory.

His date just blew into town. She's all fluttery in this gaudy multipart
outfit she copped at various exotic bazaars and flea markets. Half the

ment digests its
over-rehearsed rhinoceros. Bookworms excrete monogamous bunnies.
Blue crud excites red ecstasy. All this during the furious sleeping of
colorless green ideas.

The Ghazal Gene

The ghazal is an old poetic form, very old, very stringent, very strange.
It is older than the sonnet. Or so writes the poet Agha Shahid Ali.
According to Ali, ghazal is pronounced to rhyme with muzzle, and the
initial *gh* sound comes from deep in the throat like a French rolled r.
Like a smoker quietly clearing his throat.

The ghazal goes back to seventh-century Arabia, perhaps earlier, in
contrast to the sonnet, which goes back to thirteenth-century Italy. If
grammar is genomic, could the ghazal be genomic?

A ghazal performs itself in couplets, five or more. The couplets have
nothing to do with one another, except for a formal unity derived from
a strict rhyme and repetition pattern.

In the last couplet it is customary for the poet to mention him or
herself by name, by pseudonym, or as "I." In all other couplets this is
strictly illegal.

The ghazal is the form of choice for the incorrigible narcissist
because it always returns to the subject of the poet, rather like a social
bore at a cocktail party.

The ghazal in English has been tortured and butchered, and this
pained Agha Shahid Ali and moved him to write a rant. This head-
strong but humorous harangue precedes an anthology of good ghazals
in English, *Ravishing DisUnities*.

Or maybe they're not so good. Some are exquisite. Others stand in complete violation of Ali's ground rules. What does it matter? Who cares?

If you construct a ghazal on a subject so that each couplet chews on the theme announced in the title like a meat chopper, or if you violate the form by using slant rhyme—say white/what instead of white/fight—or if you violate the rule of no enjambment between couplets—the form disintegrates. The eerie magic of the ghazal, its ravishing disunity, its weird indirection—falls to pieces. The thing becomes awkward, stiff, forced like a too-fancy, out-of-date party dress purchased at a thrift shop that besides missing a button, is too tight—unsightly.

I have committed god-awful ghazals. At first, I missed the point about autonomy of the couplets. Then one day I was visited by the muse, Keeper of Classical Forms. Perhaps she was sent by Agha Shahid Ali, who died of a brain tumor on December 8, 2001. He was fifty-two years old.

I gutted my ghazals and began again.

Genome Ghazal

One earth, one ur-gene, in the beginning.
Mountain air. No green, in the beginning.

Black towers. Steel and glass. Blue dawn
downtown. Pristine in the beginning.

Old friend, did you slip into not-being,
or was death like a dream, in the beginning?

Dirt-obliterated bones, bits of bowls,
stone tools—unseen in the beginning.

Sibilant hiss, susurrus sigh—Priscilla—
What did it mean, in the beginning?

In the Beginning

When I was twelve, I took up bird watching. On the first day of my new hobby, I set out down the dirt road of our dairy farm noting in my tablet any bird I saw. Crow. Red-winged blackbird. Sparrow—I had no idea what kind. Turkey buzzards spiraling down. A cardinal flashing red in a black locust tree. That evening over supper, I read my list to my brother and sisters, and to my rather worn-down parents.

The next day my twin sister, Pammy, took up bird watching. She returned with a list twice as long. Besides my birds, she had recorded a wood thrush, a black-capped chickadee, and a yellow finch. Our mother put an immediate stop to Pammy's bird watching hobby. She forbade Pammy to watch for birds or to put down the names of birds. Pammy was not even to speak of birds. Bird watching was my hobby, not Pammy's hobby.

Pamela and I each, like everybody else, have three trillion cells, give or take a few. Most of these cells have at their center a copy of our genome. My genome is identical to Pamela's genome. Therefore, Pamela and I feel we have something to interject into the debate on cloning. But here I speak for myself.

I speak for myself because I am looking out of my own eyes. I live in the Puget Sound region—a land of clouds, salmon, Orca whales, congested traffic, and double-leaved bascule bridges. Like many Seattleites, I grumble at the excessive sunshine in mid-July. I like foghorns and ducks and snow-capped mountains. Rainy Seattle with its cafés and bookstores is a perfect reading-and-writing city and I am happy here, happy as a coot bobbing on Green Lake. My place—the Pacific Northwest—affects who I am.

Genes don't even determine all physical characteristics. I have curly hair; Pamela has straight hair. This could be the weather, or maybe I have more kinky thoughts.

Once it happened that an old friend of mine, long out of contact, saw Pamela in Washington, D.C., jogging in Rock Creek Park.

"Priscilla!" she screamed.

"I'm not Priscilla!" Pamela called back. She waved, but did not bother to stop.

Years later I reunited with my friend and she informed me of my mental lapse, my rudeness, my inexplicable behavior. I reminded her that I have a twin sister who may or may not have identical fingerprints. In any case, I'm not responsible—for anything.

In my memory, our childhood is fused. For years I told the story of how our mother taught us to read at the age of three. Once I told the story in the presence of my mother, and she informed me that Pamela had learned to read at the age of three. I had exhibited zero interest in reading until I was six or seven. I must have thought, as Pammy was learning to read: Oh! Look! We can read!

Twins share the same genome, but they do not share the same environment. One twin dominates; the other carves a niche out of whatever space the dominant twin—in our case Pamela—leaves available. One may be more conservative, the other more deviant.

Our desires send us out on our various paths; they color the persons we become. Pamela grew up wanting to be a scientist, and at eight or nine this moved her to collect white mice and to experiment with questionable liquid mixtures in her chemistry laboratory. When she was sixteen (in the bad old days of 1959), she wrote to medical schools asking how she should prepare herself to be admitted. Each and every school wrote back: girls need not apply. We are formed by our generation, our era, as much as by our genes.

But times changed. After Pamela graduated from college and worked for a decade as a social worker, she came to her senses and got a PhD. She is now a brilliant historian of Renaissance science and technology.

I wanted to be a poet, and that sent me down a different road.

If a twin is not the same person, why would a clone be the same person? How could you replace one twin with another? Each looks at the world through his or her own eyes. Place, choice, chance—all affect who a person is. Who could imagine that one person—that ineffable, multivarious complicated constantly changing complexity that is a single human being—could be the same as another?

Today Pamela and I are the best of friends, a mutual aid society, career consultants, fashion consultants. I live by myself; she lives with her husband and often visits her daughter, son-in-law and grandsons. We both write books—utterly different sorts of books.

I'm not a bird watcher, but I like watching widgeons paddling about on Green Lake squeaking like a flock of bathtub toys. They look identical to me, probably because I do not take the time to distinguish their particulars. Pamela would do better. I think she has a Life List, and I think widgeons are on it.

Dolly

Dolly, cloned from an udder cell of a six-year-old sheep, was born on July 5, 1996. She looked very lamblike, with her white wool and curious eyes. Dolly the newborn had six-year-old cells. She soon went stiff with arthritis. She soon came down with lung disease. Sheep live for eleven or twelve years and in old age typically suffer arthritis and lung disease. Dolly's caretakers, considering her progressive lung disease, put her to sleep in February 2003. She was seven years old.

Dolly illustrates the difficulties of reproductive cloning. She was just a lamb, like any other lamb, soft and woolly and frisky. But she was one cloning success out of hundreds of failed tries, and even then, she had complications and died young, if you count her age from the time she was born. Since Dolly, other large mammals have been cloned. One calf's hind-end is fused into one back leg. Extreme abnormalities in cloned animals are routine. Life is not easy to create in the lab.

The idea of using reproductive cloning to clone human babies is fought, and it's fraught with the nightmare of grotesque "successes"— infants with severe abnormalities. Any cloned infant will enter a life of many problems and early death. The most heartwarming argument used in favor of reproductive cloning is that human cloning could provide the grief-stricken parents of terminally ill babies a copy of their lost child. It could give them their baby back.

I am here to speak as one of nature's clones. A genetically identical being is not the same being. A cloned baby would not return a dying baby to its parents. It would not delete the grief of losing a child. A cloned baby is a different baby. It is an identical twin, not the same little boy, not the same little girl. A cloned baby would start life in the wake of grief and death—already a vitally different life-beginning. It would delete neither the death nor grief over the death of the child that lived for only a short while. Imagining that a cloned baby could replace a lost child is as insensitive as the idiot-persons who say to grieving parents, "You can always have another child!"

Stem Cell Research

But stem cell research, that's a different thing. Stem cells are fetal cells, but no born child is involved.[1] Stem cells are the body's ur-cells, the first to grow after the sperm and egg join. Stem cells are poised to become any body tissue, from liver to brain to skin. Stem cell research holds the promise of curing paralysis, Alzheimer's, Multiple Sclerosis, Parkinson's . . .

To my way of thinking, stem cells are not a human being but a potential human being. I do not disrespect the Right to Lifers, but I've always wondered why they don't join the increasingly effective campaigns to save the infants and toddlers who die every year worldwide (in 2013, 6.5 million deaths of children under five)—preventable deaths of born children.

The Ancient One

Looking into this petri dish, into this dish of our own cells, we can see, after a fashion, our ancestors. We can unravel their journeys. It's as if DNA were a telescope with multiple lenses pointed at the deep past, each lens revealing a different scene. The Human Genome Project, added to the archaeological breakthrough of carbon dating, added to new archaeological digs, added to the study of languages living and dead, added to the study of blood types, added to sonar sweeps of ocean floors that were once dry land will rewrite the story of who we are and who our ancestors were.

We know now that *Homo sapiens* spread out from Africa. That is a long story. We know the species spread to Asia and to Europe. Another long story. Then some of them came to America.

1. The recent discovery that stem cells can be produced from adult cells presents the possibility of going forward with stem-cell research while avoiding the whole embryonic-stem-cell controversy.

The old story is that peoples out of some sort of Asian gene pool walked to the North American continent over the Bering land bridge, when the Bering Strait was iced over, some twelve to thirteen thousand years ago. These people, these ancient ones, evolved into American Indians, into South American Indians, into Cherokee and Crow and Sioux and Mayan. That's the old story. A newer old story is that they came earlier, in waves, and that some may have come by boat.

Kennewick Man threatened to rewrite the old story. Teenagers found a man's bones half-buried in a bank of the Columbia River, in eastern Washington, during Kennewick's annual unlimited hydroplane races on July 28, 1996. The bones were determined to be 8,500 years old, one of the oldest complete skeletons ever found in the Americas. Controversy flared when an archaeologist working for the Benton County coroner's office declared that they were Caucasoid bones (a white man's bones). The skeleton had a narrow, elongated skull, like Europeans, unlike Native Americans. This would suggest that the ancestors of Europeans arrived in the Americas before the ancestors of Native Americans did.

However, genetic research has uncovered that Native Americans have a common ancestor with native peoples who now occupy south-central Asia. Several of these peoples have narrow, elongated skulls. The scientist who breezily declared in the first week of the find that Kennewick Man's bones were white man's bones spoke in haste.

In the case between Native Americans who want to bury the Ancient One's old bones on the one hand, and certain scientists on the other, the courts ruled that Kennewick Man's bones may be studied. In 2015, after many unsuccessful efforts, Kennewick Man's DNA was successfully extracted. Yes, he was a Native American. Yes, he was closely related to one of the tribes—the Colville Tribe—that claimed him as its Ancient One.

My Ancient Ones

My ancestors *are* European. They came out of Africa, just as all of our ancestors did. They lived for generations in the Near East and in eastern Europe. There they met up with some Neandertals, and whatever happened between them gave me my Neandertal genes. During these many generations, groups that would become Asian were moving east, probably along the seacoast. Eventually some arrived in North America. At the same time, groups that would become Caucasian were moving into what were then the steppes of Germany. The earth was becoming colder. *Homo sapiens* were hunting with more social cooperation than before. Neanderthal bones show that the Neanderthals were having a hard time. They were starving. The last ice age lasted a long time. Then it got warm again. Germany grew trees. Germany grew the black forest. Children played in the woods, got lost in the woods. The woodcutter's children, Hansel and Gretel, found their witch . . .

The Courage of the Ancestors

Back in the forested hills and hollows of Old Germany, the Brothers Grimm went about collecting stories—fairy tales, legends, riddles, ridiculous superstitions. This was in the early 1800s, but the stories they collected were of course much older, handed down from previous generations. Grimm's fairy tales are known the world over and can be compared to analogous fairy tales from just about every culture. Their legends are less well known. One of them, Number 328 in the Brothers Grimm published collection, is titled "The Dead From the Graves Repel the Enemy."

According to this legend, the town Wehrstadt got its name—related to the verb "wehren," to repel—following this event. The town suffered an attack by "foreign heathens" of vastly superior force. At the moment of defeat, the dead rose from their graves "and courageously repulsed the enemy, thus saving their descendants."

Grandma Henry's Love Story

My mother's father, Robert Bauman Henry, whom I called Granddaddy, was hard working and rather taciturn. He spoke little, except when he was laughing and talking in Pennsylvania Dutch with his insurance customers. My mother's mother, Olive Erisman Henry, whom I called Grandma, talked in a constant stream in English, considering herself to be emancipated from Pennsylvania Dutch. My grandparents did not speak overly much to each other.

One day Grandma told me the following story. Decades after their wedding day, their three children grown, grandchildren already born, Granddaddy told Grandma, "You were the most beautiful girl in the whole town!" At this point, Grandma paused in her telling of the story. Then she said, "Why didn't he ever tell me that before? I never knew I was beautiful!"

Mother's Love Story

My mother once told me, "I was the adored first child."

My mother wrote to her mother, my Grandma Henry, every single week from the time she went away to Bucknell College to the time (that same year) she married my father, and these young people had their first three children before they turned twenty. She continued writing to her mother every week, regular as clockwork, for decades, until her mother, my Grandma Henry, died on August 29, 1987.

My mother's own dying was long and painful, involving diabetes and strokes. During the long years of her extreme disablement, my father was her caretaker. Dr. Barbara Henry Long died on May 29, 2003, at 11:45 at night.

A couple of weeks after she died, in the midst of all the turmoil and arrangements, my father took out a framed photograph of my mother, taken when she was eighteen. "She gave me this picture after our first date," he told us. In the photograph, Barbara Jane Henry is young with a long and thinner face. Her brown hair curls softly around her face and her eyes are shining with happiness.

Naming Names

The crime of Christoph Tanger, a German innkeeper, was stealing horses. He was tempted by the devil to associate with thieves. These are the facts reported in the printed account of his hanging, which took place on March 13, 1749, in Gemersheim, a town on the Rhine River in what is now southern Germany. The "leading out" of Christoph Tanger occupied four hours. The procession cheering him on to his execution sang "more than 20 of the finest Evangelical Lutheran hymns." Upon "entering the circle" it was intoned, "Now we are praying to the Holy Spirit." Christoph Tanger himself thanked the Lord and, according to his pastor, "repeatedly recommended to me his wife and children, that the latter should be raised in his religion which is so much a consolation to him. Whereupon under constant cheering up he died without much pain!"

Two years later Christoph's widow, Anna, and their children arrived in Pennsylvania. Their German became Pennsylvania German, their Dutch became Pennsylvania Dutch. I am here because of the broken love between Christoph and Anna. I am here because of their son Andreas, witness at age six to his father's broken neck. I am here because of the love between Andreas Tanger and Catherine Lottman, married in 1768. I am here because of their children and their children's children, ending with my mother. They are the vessel from which my genes were poured. They are the ancestors who gave me this world. They are the lovers who put me into this blue dawn, watching and listening . . .

III Stone

An acre of stony ground,

Where the symbolic rose can break in flower ...

YEATS, "My House"

12

Disappearances

The process we call destiny
in which we are the material to be dissolved.

AGNES MARTIN

Long before my sister Susanne vanished, I would swim at times in an uneasy dread of sudden disappearance—of persons dear to me, or beloved animals, or treasured objects. Zeppy, our rust-colored, white-ruffed collie. I was six when our family piled into a farm truck and drove out of Pennsylvania forever, without her, the wet-tongued friend of my childhood. Perhaps in the turmoil of packing she had run away. Years later, my most prized possession—a pair of white figure skates—vanished from the trunk room of Moravian Seminary where I was a fifteen-year-old scholarship student. I never skated again. Later still, in Manhattan, my roommate's boyfriend—I forget his name—one night announced his intention to go out and buy a pack of cigarettes. He never returned. It took Jane some months to work out that he was not dead, that he had used the cigarette ploy to save himself the fuss of leaving her. Now, long since married to a devoted husband, she looks back on the episode philosophically. To this day I view it with horror.

Yet, did I not once, years ago, permit a friend of a friend, a distinguished looking gentleman in tweeds and iron-gray sideburns, to hide in my Boston apartment while he arranged to vanish from his old life? He spent the first evening on nomenclature, entering potential names into a pale green tablet. The next day he went out wearing a black turtleneck, wool trousers, and his new name. He easily persuaded the naïve Catholic girl working at the Social Security office that he had toiled these many years as a missionary in Bolivia and found himself approaching middle age with no social security number. It took him exactly three days to build a set of ID, to become in one life, a missing person wanted for child support, in another, a genial ex-priest given to tweed blazers with leather-patched elbows, a Jesuit recovering from experiences in the rain forest difficult to speak of, even now.

And did I not myself once disappear; vanish out the back door of a dinner party given by dear friends? The three other guests were insufferably pompous, or so it seemed to me. I wanted desperately to be gone, so slipped into a backyard lit by a yellow moon, sped past shadow trees and hedges out to Sacramento Street. I strode home through the moon-glittered cityscape, exhilarated at my brilliant escape, without a second thought for my friends, who called the next morning to confirm my continuing existence.

But I did reappear the next day. All's well that ends well. Not so my best friend in high school. Margaret Gaines, or Peggy as we called her, was expelled from Moravian Seminary for Girls in December of our senior year (1960). She had spent the night (alone) in the chair of a hotel lobby while another girl went upstairs with her boyfriend. (The other girl was also expelled.) We were all, of course, applying to college. Expulsion, for Peggy, brought an immediate halt to all the anxious and exciting preparations for crossing the great divide between high school and college.

Moravian Seminary for Girls. A brick and stone manor house set in terraced gardens running down to Green Pond, near Bethlehem, Pennsylvania. A private estate turned boarding school. It was a good school, but also rigid, religious, rule-ridden. The headmistress, Miss Lillie Turman, was obsessed by appearances, notwithstanding the school motto, *Esse Quam Videri*, To Be Rather Than To Seem.

Peggy Gaines was from Manhattan. When we first met—we were fifteen—I was deeply impressed to learn that a taxicab had transported her from New York City to Moravian Seminary, for the small fortune of one hundred dollars. The Yellow Cab was loaded up with the equipage of a boarding school girl of the late 1950s—school uniforms, bongo drums, a guitar, a record player, phonograph records, Dial soap, Prell shampoo, pink flannel pajamas, pleated skirts, Oxford button-down shirts, books, notebooks, stationery, pens and pencils, three-ring binders, a tennis racket, ice skates, saddle shoes, pumps, penny loafers, slippers. The driver had conveyed these objects, meter running, along with Peggy herself from the Big Apple to the rolling hills of eastern Pennsylvania, a distance of two hundred miles. He stopped for a snack whenever Peggy directed him to do so.

It is remarkable how little I remember about our friendship. I know we talked in a seemingly endless stream throughout the tenth, eleventh, and twelfth grades until the conversation was terminated by her expulsion. In her last letter to me she wrote, "All our ideas were wrong." I have the letter, handwritten in dark-blue ink on crinkled pale-blue stationery folded into a crinkled pale-blue envelope.

What ideas? We neither dated nor did drugs. We were good students. The most radical thing we did was listen to the Kingston Trio—I remember the "Sloop John B." We declaimed the beat poets. We read Gregory Corso's "Bomb" to the dum dedum of Peggy's bongo drum in the flicker of a smoking candle. I adored the lines of a Lawrence Ferlinghetti poem that went: "I am awaiting perpetually and forever a rebirth of wonder." We were devotees of the Hindu poet Rabindranath Tagore.

Were we boy crazy? I don't think so. We considered ourselves sophisticated and intellectual. Like all sophisticates, we smoked Benson & Hedges cigarettes holed up somewhere in the warren of cellar rooms that underlay the manor house. One of these rooms, alas, was designed to convey steam heat to the wainscoted and velvet-curtained interiors above. Cigarette smoke wafted up among the steam, and for the crime of smoking cigarettes I was suspended for a week. But Peggy was never caught smoking.

In my reply to Peggy's crinkled blue assertion that all our ideas were wrong, I wrote that we would always be friends.

I was wrong. I never heard from her again.

Peggy was tall with a dark mass of curly hair and clear white skin given color by a light sprinkling of freckles. (I could be making up the freckles.) She had a strong-boned handsome face. Out of uniform, she wore a black watch–plaid pleated skirt, a pale-blue Oxford-cloth button-down shirt pinned with a gold circle pin just below the right collar, and penny loafers. Whenever a fellow student farted audibly, Peggy would wave her hand kindly and say, "Natural human function!" She was an only child who lived with her mother. I do not remember anything about her father.

I sometimes try to work out why she vacated our friendship. I suspect it became fused in her mind with the painful expulsion. Possibly everything associated with Moravian Seminary, including me, turned bitter. Perhaps she blamed me for inciting her to sit overnight in a hotel lobby, though I knew nothing about it. But certainly, I questioned every rule. I walked barefoot in the hallways. I read the Romantic poets after lights-out, secreted in a closet under a blanket, illuminating the page with a flashlight. And of course, I smoked tobacco cigarettes. These were the crimes of my youth. Possibly I became fused in her mind with a part of herself that she now rejected—she was an adventurous, questioning, intellectually alive teenager who was severely and unfairly punished for a mild offense.

Her family dealt with the disaster by sending her to a Swiss boarding school (I do not know which one) for an extra year. Perhaps, then, she walked into a European world so sophisticated and entrancing that Moravian Seminary for Girls with all its dumbness faded quickly away.

She is in the world somewhere, in Paris or London or Bangkok, on some street, in some house. She is not dead, or at least her name does not appear in the Social Security Death Index.[1] She lives somewhere in the familiar world, a world we both know and don't know, a world riddled with strangeness and mystery. She will be in her seventies, happy I hope, likely with a different name, likely with children, with grandchildren. Possibly she rejected not only me but the United States and became a citizen of Switzerland. Possibly she became a Buddhist,

---

1. Correction. There is a Margaret Gaines in the Social Security Death Index. She was born in 1943, a bad year, since that's the year both my friend and I were born. This Margaret Gaines died in Alaska in 1990, at age forty-six. When I first happened upon this Margaret Gaines, I decided it could not possibly be my old friend. I did not even consider it. I was in complete denial. Now I'm not so sure.

or a Catholic. Perhaps she is a medieval scholar who divides her time between France and Florence. Or a medical doctor in a white coat. Or an anthropologist. Margaret Gaines, girl of many talents, many potentials, a brilliant girl, sophisticated for her age, the daughter of a cultured family of theatergoers, film connoisseurs, museum patrons.

I have occasionally taken a stab at trying to find her, without success. She has not looked for me, or so I must assume. I'm easy to find, a writer with the same old name I've always had. So be it.

The pain I felt at her loss was excruciating. I remember returning to school after the Christmas break, standing on the grand staircase, its red carpet, its polished banister and newel post, its white balustrade—feeling her absence as if she were my amputated right arm. As for Moravian Seminary, the school had purged itself of its disgraced students. Not a word about them was ever spoken again. They were simply gone.

A disappearance is not a hard fact, like bankruptcy or war. It is a circumstance, a situation, a seeker's perspective. Where are my glasses?! The object has vacated, not the world but my field of vision. There are eternal mysteries, the mystery of the missing sock. There are so many missing socks that we might almost expect one day to come upon Sock Mountain, where generations of socks gather and accumulate. Margaret Gaines did not disappear from herself, nor from her mother, nor from the sun-warmed, glittering world. She is lost, but only to me.

For me, her disappearance no doubt stirred up an old disturbance lingering from the time my mother and twin sister vanished. I was three. My parents drove me to the distant Philadelphia Children's Hospital for an operation to correct my crossed eyes. In accordance with pediatric customs of the 1940s, they gave me over to the care of nurses, returning to pick me up ten days later.

They were gone. I was alone with strangers. White-dressed nurses with squeaky shoes. On the operating table I dreamt the walls opening and trucks driving in. Afterward, my eyes were bandaged. The nurse or nurses fed me, intoning, "These are peas," "These are mashed potatoes."

Ten days later when Mummy returned, I saw her as a stranger through a glass partition, wearing a skirt, carrying a brown suitcase. I felt nothing. I did not recognize her. She was no one I knew, or wanted to know. Too much time had passed. Our love affair was over.

At home, I carried my shoebox stuffed with get-well cards upon my person at all times. I reviewed my cards at every opportunity, ordering them, reordering them, gazing at them in wonderment. There were blue cards, red cards, purple cards. Rosy-cheeked grandmas, cheerful mummies, red-smiling clowns carrying red balloons, yellow-pigtailed girls like myself, getting well soon. Puppies and kittens and black-and-white Dalmatian fire-engine dogs. All brightly wishing me love and a speedy recovery. One day I was reviewing my treasures on the sun-whitened patio, when I sensed my mother looming at the screen door. She swooped down like a hawk and snatched my shoebox of cards—this happened in one split second—and I never saw them again.

From the molecular or atomic point of view, the brightly colored cards remain in the universe, mingling with their comrade molecules. Matter dissipates but does not disappear. I myself forgot about them for a long time. Then one day as a college student I remembered them, and bought for myself a bright silk blouse—blue, red, and purple—to stand in for my shoebox of get-well cards. The bright cards lived on in my subconscious, emblems of love that glowed like coals within me. They still exist in my imagination, and in the stuffed shoebox that is memory. The mystics say that everything is transitory, that everything that comes, goes. I say that, one way or another, you get to keep all the love that has ever come your way.

Susanne was forty years old when she vanished. I was forty-three. It happened during my summer as a history scholar at Harvard University's Bunting Institute in Cambridge. From my new home in San Diego, I had returned to my old community, my old stamping ground.

Boston/Cambridge was replete with old friends who that summer became recruits to the search party for my lost sister.

Before Susanne disappeared, I was living happily in my friend Joan Mandell's furnished apartment in Porter Square, which she rented for the use of friends and for her own visits. It was a three-room railroad with wooden floors, throw rugs, pillows, couches, window seats, simple chairs, simple curtains, a place kept cheap in that skyrocketing housing market by the perpetual odor of McDonald's grease wafting in night and day from the nearby fast food venue. I didn't mind it.

I worked at my very own office at the Bunting Institute every day, and in the evening indulged my research obsessions at one or another of the one hundred Harvard libraries. Each evening I'd return to my comfortable digs about 9 or 10 o'clock pleased with my work on the technology chapter of my book on the history of coal mining. I'd get my phone messages for the day.

It was a balmy summer night on July 21, 1986, when I came in and cheerfully clicked on the answering machine and heard my mother's cracked voice: "Susanne has run away."

Susanne was an ethereal-looking, high cheeked, wide-mouthed, gorgeous woman whose astonishing beauty—equal to that of Queen Nefertiti—used to make people gasp. With her long fingers, her Grecian nose, her high forehead, she had dropped into our family of stubby fingers and peasant noses apparently from nowhere, unless you look at a young picture of our grandfather Walter Long, writer.

Susanne played the recorder. She had a master's degree from Georgetown University. She taught English as a Second Language to Cambodian refugees. She was a watercolorist, a painter, a craftswoman who batiked fabrics and stitched patchwork quilts. She loved to walk in the woods, to sit by the sea. She had spent a year in Africa, four years in Morocco. She spoke Arabic, and once, in the old, crowded, holy city of Fez, my mother witnessed her turn to face a group of Moroccan men who had apparently mouthed obscenities to this tall, fair American. She cursed them out in loud, rapid Arabic. To judge by their shocked retreat, she had a good technique going.

She had straight, thick, fine straw-colored hair, rose-and-porcelain skin, and exquisite taste in clothes and in jewelry, another anomaly in a family frumpy with hand-me-downs. Her eyes were sapphire blue, gemlike, and they flashed like eyes in a novel. She had a funny sense of humor, and people were drawn to her. I was drawn to her. She was three years younger, and when I was about twenty, we became fast friends, giggling on and on about our mutual old ladyhood, which we planned to spend rocking on some creaking front porch with our many cats and dogs, sipping an excellent red wine, commenting on the scene.

Her voices began when she was thirty-two or thirty-three, threatening whispers, at first seldom. She had married the love of her life, and now got a divorce. Gradually, in the midst of a teaching career, in the midst of sumi-e painting, in the midst of a move to Seattle, to its ducks, coots, cafés, cappuccinos, in the midst of preparing lessons and going for walks around Green Lake, the voices became more insistent.

So began the long years of trying to save her. She was stubborn as a mine mule. She detested Haldol. She denied she was crazy. On Haldol, she said, you don't feel anything. She would rather be dead than live on Haldol.

Once, off Haldol, she became furious at me when I failed to perceive the entire Red Chinese Army bombarding the windows of her garden apartment on Capitol Hill. What was I doing? Here we had an emergency, and I was sitting with my nose in a book!

At the time of her disappearance, she was living voluntarily in a mental-health facility near our parents' house on the Eastern Shore of Maryland. My father had taken her to lunch. They had returned and, at the front desk, signed her back in, as required. When, six hours later, my mother arrived to take her to dinner, she was gone. Hours earlier, the caretakers had noted her absence, conducted a cursory search in the nearby dense woods, and concluded, notwithstanding incontrovertible contrary evidence in the book at the front desk, that she and my father had not returned from lunch. Thus was six hours lost before she was officially designated a missing person. Time to get to the Greyhound station. Time to hit the road, to take off, to make tracks, to wish the world goodbye.

My father searched the dense woods near the mental-health facility with his dogs. He did not find her.

A drowned person stays underwater for three days, then rises to the surface. In the wetlands of the Chesapeake Bay, the salt marshes, creeks, tributaries, in the wide Chesapeake itself or in the Chester River beside whose brown waters we'd played out our childhood, Susanne might easily have drowned. The sheriff explained about the three-day rule for drownings. We waited out the three days. Day one. Day two. Day three. No Susanne.

We had the thought, never followed through, that we should engage professional dogs from Baltimore.

Back in Cambridge, I had been invited to dinner by my friends Louis Kampf and Ellen Cantarow. I called them to explain that I might bring Susanne. I was that sure she was on her way to Boston.

The first immediate task was to make a missing person poster. I used a picture of Susanne's Haldol-swollen face, what she looked like now. The poster asked, Have you seen Susanne? It gave a description, including VERY BLUE EYES.

The world went about its business as usual. I became completely preoccupied with the effort not to burst into tears at the Bunting Institute or in the library. I continued working on my book during the day and devoted my evenings to the study of how to find a missing person. The kindness of people at the National Center for Missing Persons, whose own persons were missing. The kindness of people at the homeless shelters. Kindly police officers and kindly sheriff's deputies. Kindly telephone operators who found telephone numbers and said they were sorry.

One evening, I returned to my Porter Square, McDonald's ambience apartment, and it began to rain. It rained and rained, a drenching Boston summer rain. In the solitude of my apartment, with rain drumming the roof, I began sobbing at the thought of Susanne lost out in the rain. I sobbed all evening and fell to sleep sobbing and the next morning woke up still sobbing. I simply could not stop. I dressed and brushed my teeth and made my way to my office at the Bunting Institute, continuing to quietly sob.

I went to dinner at Louis and Ellen's without Susanne. The search party/Boston Division pasted missing-person flyers on every lamppost, blank wall, and telephone pole in Boston and Cambridge. Susanne's face was also plastered all over Seattle, all over Washington, D.C.

The summer ended. I returned to San Diego.

Again my father searched the dense woods with his dogs.

The thought began percolating among some members of the family that she was no longer living. If she was lost in the woods, it was said, she would be found in November when deer-hunting season opened.

In San Diego, a schoolteacher I knew told me the story of her brother. He was twenty years old, and one day took the train from New Jersey to Manhattan. He was never heard from again. That was forty years ago. But Susanne began being sighted in homeless shelters (where the Search Team continued to plaster posters). She was sighted in Maine. She was sighted in Worcester. She was sighted in a homeless shelter on Cape Cod. The calls kept coming in. It was only a matter of time before she would be found.

Then she was found. She was found on the Eastern Shore of Maryland in the dense woods my father had searched at least twice with his dogs. On November 7, 1986, two hunters deep in these woods shot a white-tailed deer. They were chasing it down and came upon what they thought was the doe on the ground, her white belly. They looked closer and perceived the white bones of a human skeleton. They went to get the sheriff. The medical examiner positively identified my sister, Susanne Long, through dental records, before we were informed. Case closed.

Her last words to me: How's your writing going? A fellow sufferer from mental illness had witnessed her waving goodbye to my father, after lunch. According to this person, she had given him a dazzling, almost angelic smile. Gradually we pieced together that to every dear person in her life she had said goodbye before she walked into the woods and out of our lives forever.

I think of my sister's bones lying clean and white in the woods. It is a dense wood, low and flat, with greenbriers tangling through sweet gum, oak, and hickory. I think of the high flush of her skin—gray and maggot-eaten. After death, the body's resident bacteria begin feeding upon it. Bacteria create the putrid smell that attracts bluebottle flies from miles away. Fly eggs hatch a pudding of white maggots feeding on rotting flesh. Frogs croak nearby, for nothing on the Eastern Shore is far from the salt marshes and creeks and inlets of the Chesapeake Bay.

Growing up in these wetlands, we searched for lost cows by looking skyward, at turkey buzzards spiraling down into smaller and smaller circles toward the dead animal.

I see the descent of the death-buzzards toward my sister's flesh. I hear the flapping confabulations of crows hopping about the feast. The banquet will have created a racket—croaks, caws, buzzings, hummings. Perhaps a birdsong makes a melody in the near distance. The cacophony of celebrants gather about her remains, the remains of her astonishing beauty, the remains of her four decades, the remains of her startling blue eyes, the remains of her long fingers, her recorder music, her love of Tolstoy, her large, high-arched feet.

I imagine her bones white as the belly of a white-tailed deer, her eyes hollow sockets, her teeth, even and ivory-colored, longer without their wide mouth. I think of her hollow cheekbones, more hollow than before.

I imagine small grackles and sparrows hunting seeds among her bones, pecking whatever is left of her slenderness. It seems to me as I grow away from her death, as I view it now from a distance of years, that her bones lying lost in the woods were like a small altar, forsaken by church and crone, smokeless, dechaliced, whitening among grackle and wood thrush and flame-colored autumn leaf. The cloistered life of the thicket—motley, croaking, fervent—made its way to the sacral wreck, paid homage in droves and swarms and flocks until it was consecrated, completely. Amen.

13

Archaeology of Childhood

Each one of us, then, should speak of his roads, his crossroads,
his roadside benches; each one of us should make a surveyor's map
of his lost fields and meadows.

GASTON BACHELARD

Introduction

The research reported here is directed toward the study of the remains
of a human childhood that occurred on the Eastern Shore of Maryland.
The site is situated on the Chester River at Comegys Bight, a mile-wide
bend of the river. The Chester River is the ancestral village site of the
Algonquin-speaking Wicomiss Indians, a people exterminated by
English colonists in the Wicomiss War of 1669. The river is a tributary
of the Chesapeake Bay, the largest estuary in the United States—in geo-
logic terms, the drowned ancestral valley of the Susquehanna River.

The house of childhood under study—that of "twin" or "Poky"—
stands at the end of a deeply rutted and presently impassable dirt road
that wends for a mile through greenbrier-tangled woods, the marsh,
abandoned pastures. It was occupied ca. 50 BP.

The researcher and collaborators returned to the site on foot on the
afternoon of December 26, 1998, tramping over the snow-tracks of red
fox and white-tailed deer. Wind rattled the sweet gum husks high in the
sweet gum trees. At the end of the road, the farmhouse came into view.
It was found to be overgrown and swallowed up by Virginia creeper.

There was a rusted windmill. Sheds. Caved-in barns. Blackberry vines tangled in the doorway of the milk house. There was a wooden hut with a cement floor, "the oldest house on the Eastern Shore," at one time inhabited by African slaves.

Artifactual Data

Grackle. Black snake. The Attic.
Calf-bucket nipple.
Rubber farmboot.
Yellowjackets.
Greenbrier, Barbed wire.
Ashes and bones, my sister Susie's bones.

Structural Remains

Woody vines shutter the rooms of childhood. The brick fireplace in the kitchen was built in slavery days. The dining room—cracked linoleum, cracked fireplace, crumbling yellow walls.

We go up the creaking stairs to the landing. There is the newel post we once hung upon. There is the landing window, looking out on the long dirt lane. We—the Three Big Kids and Susie—waited there on the landing for our baby sister to come home. At the head of the stairs, the bathroom—rusty bathtub, rusty sink, rusty toilet, the ten-foot black snake in the toilet that made Grandma scream.

Virginia creeper creeps into the house, creeps along the windowsills, creeps down to the floor and along the floorboards.

Dark tiny rooms. Andy's room, Mummy and Daddy's room, Lizard's room. The attic, the domain of the twins Pammy and Poky, their whispers and dolls.

When our baby Lizard came home, Susie, age six, began to teach her to speak. It took her two years of daily work, but at last she succeeded.

Susie, the one without a room, the one without a twin, the one now dead.

Susie: The Third Twin

She lies in the darkness, away from the voices. She is a white form, covered in a white sheet. Voices reach her, whispering in the dark. She whispers to the night, to the phantom that is her twin. The twins whisper about her whispering. Whispering mingles with whispering like mist curling above the river at night, mingling with the ghost shapes of swans. Silence is Susie's music. Her sadness. Silence white as her white bones.

Locality: The Story of Miss Bell

Chestertown Elementary School was the white school. The white school was a brick building, dark inside, with classrooms and cloakrooms and wide hallways with narrow board floors. Classrooms with blackboards and chalk and high windows and rows of iron-legged desks with wooden desktops. The desktops were hinged to lift like lids to a compartment for tablets and pencils and pencil cases.

When Pammy and Pokey first moved to Chestertown and entered the second grade, all the other children crowded around them because they were the only twins. They felt pleased to be so popular, but they were shy and it wasn't long before interest fell off.

Miss Bell was old and she had tight gray curls. Miss Bell called the twins Twin. Miss Bell taught that the slave masters were kind to the slaves. She taught that the slaves would not have been able to care for themselves, but for the kindness of the slave masters.

Structural Remains

Gutter. Heat vent.
Bulk tank, barn, gate.
Windmill. Water trough.
Tool shed. Machine shed.
Tractor, Conveyor, Rust.
Rusty Pipes. Rust.

Primary Phase I: Their Happiness

We had everything we wanted on the farm—ink made of inkberries growing in great droops off the inkberry bush; and a hundred cows, sixty milking, the Holsteins with their big udder-bags that gave great buckets of milk and the Guernseys with their small udder-bags that yellowed the milk with buttermilk; and soft baby dolls with porcelain heads you could tilt backward to make them say "Whaaaa"; and a leaky rowboat with a bucket to bail it out; and books with yellow buckram covers with The Five Little Peppers embossed on the front; and Grandma-stitched black and yellow gingham dresses with yellow piping on puff sleeves; and fields of yellow flowers—buttercups and ragweed and dandelion and columbine; and our own clubhouse made from an old chicken house; and chocolate fudge cake with chocolate pudding glistening inside—Tizzy Lish Cake.

Cultural Items Not Directly Associated with Life Sustaining or Economic Pursuits

Five Little Peppers and How They Grew
Sears Roebuck catalog
"Jesus loves me, this I know."
Montgomery Ward catalog
Monopoly
Things To Do Closet
Salt Water Taffy
Cherry Ames, Student Nurse
Oliver Twist
"Old Black Joe"

Locality: The Lady at the End of the Lane

Pammy and Poky are in the third grade and the class is seeing who can sell the most magazine subscriptions. The pupil who sells the most will get a prize. Poky asks her mother if they can buy a subscription.

Her mother says don't be ridiculous they are too poor for such things. Then she tells the twins they are not allowed to go to Johnsontown to sell subscriptions. Johnsontown is the black people's village at the end of the lane. The mother says the people in Johnsontown can't afford it any more than we can.

One Saturday after her barn work is done, Poky takes her folder, sneaks out of the house, and walks down the lane to Johnsontown. At the end of the lane she looks at the neat wooden houses and decides to go to the one that doesn't have a dog. She walks up to the side door and knocks. A kindly looking black lady in a cotton housedress comes to the door.

Poky looks up at her. "Would you like to buy a magazine subscription?"

"Why come in, Chile, lemme see what you got," the lady says.

She smiles and Poky enters the house. The living room is pretty much like theirs, except that it is very clean. It even has the same linoleum on the floor, red flowers with swirling green leaves.

Poky opens her folder and starts telling the lady about the magazines: *Vogue, Ladies' Home Journal, Saturday Evening Post, House and Garden.*

The lady decides to try *House and Garden.* Poky carefully writes her name and address in the space. Then the lady asks how much it will cost. Poky looks on the chart and tells her.

Then she looks up.

The lady is blinking and her mouth is quivering. "Oh my," she says. "Oh my."

Poky doesn't know what to do. Finally, the lady wipes her eyes and goes to her pocketbook. She fumbles in it and takes out some money. As Poky takes the money she can feel the lady's hand trembling.

"Thank you," Poky says.

Then the lady asks, "How soon do it come, Chile?"

Poky looks at her chart. "In five months," she says.

"Five months?" The lady is still blinking and now her face is quivering all over. Poky is afraid she will burst into tears.

"Thank you," Poky says politely, and goes out the door.

Instead of going to the next house on Johnsontown Road, she sneaks back down the lane, past Neil Lindsey's pig, past the far cow pasture, past the marsh croaking with frogs, past the near cow pasture, and back to the farmhouse. She tells no one what she has done, not even Pammy.

Faunal Remains

Fireflies
Box turtle, white-tailed deer
Dogs: Zeppy, Laddie, Peggy, Bo, Robbie, Meg, Princess; Lady, Prince
Thumper the cat
Barn cats: William Shakespeare, Louisa May Alcott, Oliver Twist, et al.
Pussae (Latin for Puss)
Pammy's white mice
Mumbo the Elephant
Pammy's sheep*

* Note on Pammy's sheep: Pammy owned seven sheep, and they grazed with the cows and identified as cows. At the Kent County Fair, these seven sheep remained perfectly indifferent to other sheep, but they bleated pathetically at the sight of any cow.

Effigy of the Father

Winslow Long is all bone, sinew, and gnarl. Weather has rusted his scalp and his thinning hair to the colors of a wood thrush. He goes about hatless, and when he speaks he strokes his brow with the stub-fingers of a man who has worked the fields his whole life. He is a man of few words and strong values. He abominates television. He is friend to yellow finches, black snakes, crickets. He lacks certain experiences common to American life; possibly he has never been shopping. He wears baggy brown pants and plaid flannel shirts that may be twenty or thirty years old. He is beekeeper, bookkeeper, dairyman, an erudite amateur botanist. He is never without Prince, the largest and youngest in a long line of friendly German Shepherds. Recently he was appointed conservator to a large marsh. "I went," he wrote to one of his daughters, "and I found a paradise."

Photograph of the Site

A dirt lane stretches through woods of towering trees—sweet gum, persimmon, oak, hickory. Greenbriers tangle among the trunks, and crows squawk high in the branches. A wood thrush startles in the underbrush. Here is a pasture and here are two white-tailed deer bounding away. Honeysuckle tangles on a chicken-wire fence, a red-winged blackbird clucks on a fencepost. Yellow jackets buzz in the ditch. Cow fields. Daddy's row of white beehives. The herd is grazing—Holsteins, Guernseys, Jerseys—and Virginia, the draft horse that the twins ride, and Pammy's seven sheep. Soon you reach the milking barn, the calf barn, the machine shed, the windmill, the creek. You reach the old farmhouse, the din of angry shouting.

Concentration of Fire-Cracked Rocks

Anger burns the way the sun burns. The sun burns your arms and you walk down the lane burning with shame. Between the mother and the father, anger smolders and smokes and bursts into flame. This is the day that the Lord hath made. This is the day that anger hath made. This is the firestorm. This is the rage that burns childhood down to a hot crisp.

The twins walk through fire burning with shame.

The brother Andy walks through fire burning with shame.

Susie walks through fire burning with shame.

The lives of the children smoke like black candles.

Susie holds the little one's hand.

But Susie walks alone, and she burns with shame.

The fire burns bright and hot, and the children walk in its coals, they smoke in its flames, they burn in its black furnace, they smoke and burn in the flame that made them.

Primary Phase II

Then it was summer. We took off our shoes. On the day after school let out, I went into the hot sun and walked barefoot in the grass. I went into

the field and squished my toes in new cow shit. I walked down along
the fence, along the row of red cedars, through the gate to the back field
all the way to the far creek and sat at the edge of the creek. I watched
the creek ripple slow and wide and brown in the hot sun. A dragonfly
all green and purple hovered in a grass hummock. A low tree I didn't
know the name of dipped its branch into the water. A kingfisher was
fishing from that tree. The kingfisher dived into the creek with a splash
and flew up to his branch with a silver fish in his beak. He swallowed
the fish, and I watched him perch there still as a decoy until he dove
again with another splash. I listened to the hum of bugs and felt the hot
sun on my legs, and I got to thinking how glad I was that school was out
and I didn't have to go back to the fourth grade ever again, I didn't have
to look up to Miss Russell's glaring pasty face or listen to her scolding
me for not taking a bath or for not telling the truth, and I didn't have
to watch her drag Henry by his hair to the front of the classroom. That
made me feel light and happy in the sunlight, and I picked up a stick
I saw lying there and started sweeping the stick back and forth in the
brown water just to make the water ripple more. The sky was white and
hot, and I sat for a long time and a reverent feeling came over me as
if I was in some kind of chapel or something with a vault of hot sun
and white sky and dragonflies and kingfishers, and I lay down then
in the hot sun and listened to the quiet, which made a hot humming
sound. The marsh grass tickled my bare legs and I opened my eyes to
the other shore where trees were dipping their branches into the water
and I wondered what kind of trees they were and I guessed I would
never know unless I asked Daddy because Daddy knew the names of
the trees, but I didn't want to ask Daddy so I just lay there in the hot
sun and I guess I fell asleep because when I opened my eyes there was
a snake in the water, maybe a water moccasin—Pammy would know
because Pammy knew all about snakes and snails and birds because
Pammy was going to be a scientist and she collected bird nests and
snake skins and bones and mice that reproduced into more mice. I felt
happy to be Pammy's twin, happy to lie there in the sun, so happy, like I
was in paradise. I stayed there until it started getting cool. The sun sank.
The afternoon light turned to copper and red and gold. Then I got up. I

went back along the row of red cedars, through the front gate into the front yard. I went into the farmhouse, just in time for supper.

The Clubhouse

The gray wooden two-stall outbuilding called the chicken house was lost between the farmhouse and the milking barn. Daddy kept calves in the chicken house for a while, but it was really not very satisfactory, too far from the other calves, very inconvenient. From the time of the calves living there, manure in the stalls had mounted to two or three feet (clean straw on top of the old bedding made it warm for the animals). Four feet above the floor of the front room, chicken roosts, two-by-fours, spanned one stall, a foot between them. The Three Big Kids—Andy, Pammy, and Poky—asked Daddy if they could make the chicken house into a clubhouse. Daddy said yes. He told them where to put the manure. For two weeks the Three Big Kids shoveled out the manure. They got it all cleaned out and soon they opened a museum. The roosts served as display racks for bird nests, special rocks and stones, a feather, a rusted machine part that the resident archaeologist Pammy had yet to classify. They put a bucket at the door for admission. The clubhouse/museum opened for business. Susie was the first patron. She could not afford the price of admission and was therefore not admitted.

Daddy waited until the clubhouse fell into desuetude. Then he bought fifty chickens.

Effigy of Neil Lindsey

A huge, solid, umber man with a wide face and clipped frizzy hair, he worked as a waterman in the winter and as a field hand in the summer. He wore field attire, gray overalls with a square-cut bib, buckles tarnished to a dim gray metal, and a gray T-shirt with a frayed neckband—he looked like a preacher or a burnt-umber Hercules. He was by nature a gregarious and kindly man. He sat on the tractor with his thick brown hands steady on the steering wheel, looking back to make sure the children were safe in the hay wagon before he shifted into first and released the clutch. In the hot sun, the tractor chugged the hay wagon out to the hayfield where hay bales stretched out in long rows. The crew—Neil Lindsey, Buck Washington, Pammy, Poky, and the father—worked all day bringing in the hay. They said little. They took turns driving and pitching and arranging the load. When the sun sank to a red ball on the horizon, they brought the last load in, and Neil and Buck sang—deep and low, back and forth, a chant or moan about work and trouble and tired bones and being in the Lord's hands.

The Story of the Cow

The father bought a cow at an auction in western Maryland and drove it in the truck through Baltimore, where he got caught at every red light. He drove the truck down to Annapolis and across the wide curve of the Chesapeake Bay Bridge and onto the Eastern Shore. He drove through Galena, where the twins had gone to first grade, and he drove past Ravenswood, where he had worked until it became a truck farm and where the Three Big Kids had seen him rolling around in the dust with a bad man. He drove on toward Chestertown with his new cow. He crossed the Chester River on the Chester River Bridge and took a left and drove past the old brick river houses with their secret tunnels and hidden rooms, the last stop on the underground railroad before you got to the Mason-Dixon Line a few miles to the north. He took another left and drove with his new cow seven miles down Quaker Neck Road past Lee's Gas Station, past the turn to Quaker Neck Wharf. He turned onto Johnsontown Road and then turned right and drove down the long dirt lane. When he got home he backed up the truck to the loading dock and went around to unload the cow.

The cow was not there.

The next day it came out in the *Baltimore Sun* that a large Holstein cow had attended mass at a Baltimore cathedral. There was a picture of the cow on the front page of the Metropolitan Section, entering the barnlike sanctuary, looking puzzled.

Cultural Remains: Susanne

Words on a postcard:

> July 18, 1967. Dear Poky, How are you? Found a nice Haiku.
>
> > The springtime sea:
> > all day long up-and-down,
> > up-and-down gently.
> > BUSON
>
> Please write soon. Love Susie

Life on the Site in Historical Time

My mother and father decided to go and live in Rock Hall. My mother went to work in Baltimore. I grew up and moved to Boston and got a job as a printer. I was always at work and I worked twice as hard as the other printers. I missed the cud and breath of cows and was homesick for it. To be separated from my father, from our big barn, from my doll, from the garden I weeded, from the hay in the loft, from molasses milk, from Neil singing the blues, from the hot dust on our dirt road, from the nocturnal whispers of my twin gave me a pain. I went nowhere. How should I have gone anywhere? I could barely drag myself along under the burden of my memories.

Conclusion

We leave the site through the winter woods. It is late afternoon. Tarnished copper light. Crows flash black through the trees beside the dirt lane. The sun drops and the trees turn gray as an old barn. Ruts darken. Puddles turn to silver. Night comes on.

We arrive at the car, parked at the end of the lane. We get in. We drive away. Childhood lies behind us, remote, lost in the honking of wild geese, in ruins.

14

Stonework

A stone is ingrained with geological and historical memories.

ANDY GOLDSWORTHY

We are the spiritual ones, the ones who can be spooked or saved, who can be filled with reverence, with awe. "He has come to believe," Carol Shields writes of her protagonist in *The Stone Diaries*, "that the earth's rough minerals are the signature of the spiritual, and as such can be assembled and shaped into praise and affirmation."

In Scotland, England, and Wales there exist more than nine hundred stone circles or rings constructed by Neolithic peoples some six to four thousand years ago. Stonehenge, rising from the Salisbury Plain in southern England, is the most famous, though not the most ancient. Stonehenge was constructed during three different periods. The oldest Stonehenge (3000 BCE) is a circular ditch (a "henge") with a single entrance and with cremation holes around the perimeter. A wooden sanctuary occupied the center. The second Stonehenge (2500 BCE) consisted of raised stones—large sarsens (sandstone blocks) and two circles of bluestone, four-ton boulders of dolerite moved from the Welsh mountains 250 miles away, and stood upright. In the most recent renovation, done about 2300–2200 BCE, the bluestone were taken down and replaced with sarsen stones taken from a site 20 miles to the north. They weigh twenty-five tons and average eighteen feet in height. They extend underground about ten feet, and rest on chalk bedrock. These boulders were transported and erected by Neolithic (late stone age) people who lacked the wheel and who worked with stone tools. The astronomical theory of the purpose of these megaliths has been discredited. The megaliths were sacred stones, sites of unknown rituals or festivals marked by the Midsummer solstice.

Diamonds are forever. Quartz is also forever. Granite is forever. Marble, slate, chalk, basalt, and lapis lazuli are forever. Sandstone is forever. Half of all marriages end in divorce. Marriages that do not end in divorce end in separation or in death. Death is forever.

My friend Saul flies in from Boston to visit me in Seattle. One day we walk from Wallingford to Fremont, cross the Lake Washington Ship Canal on the Fremont Bridge, then huff straight up the high hill of Queen Anne. People working in their gardens on the steep slope call out encouragement as we trudge upward. We gain the summit, walk across the hill and down the other side and on to Pike Place Market where we take seats at a French restaurant and eat steamed clams dipped in butter and French bread. We take the bus home. The next day Saul takes a ferryboat across Puget Sound, and drives west on the Olympic Peninsula to the coast, where he stands at the edge of the Pacific Ocean. He returns to Seattle with the gift of an octagonal jelly-jar full of beach stones. The pebbles glow under water in tints of purple, gray, blue, white, and jade.

Who ever said, "Sticks and stones may break your bones, but words will never hurt you"? What an idea!

My mother always used to say, "Sticks and stones may break your bones, but words will never hurt you." My mother was shaped perfectly round like a soft boulder. Her girlhood photographs show a slender young woman, but ever since I can remember, she was overweight and had difficulty moving. She moved slowly and commanded that things be brought or done while she lay on the bed or sat on the davenport. Yet she had a keen mind and a sharp tongue. In a crowd, my mother could seem to be sitting by herself like an isolated boulder on a beach, without speech.

Families can be found with this peculiar constellation of family members: high-functioning but on the autistic spectrum; schizophrenic; creative. The autistic ones are reclusive types, emotional stones, physically immobile. Yet their intelligence gives them much to offer: perhaps it is out of love for my stonelike kin that I keep stones in my house—alabaster, coal, bluestone, quartz, granite, calcite, pyrite, sodalite, hematite, slate, malachite. . . .

In the 1830s an enterprising miner dug a wagonload of anthracite coal from a pit somewhere near Wilkes-Barre, Pennsylvania, and hauled it down to old Philadelphia. Wilkes-Barre is not far from Bucks County, where much later I was born to a Pennsylvania Dutch mother and a half-Scottish father. Back in the 1830s the enterprising miner got his horse and wagon to Philadelphia and sold the hard coal, but the purchasers could not manage to ignite it. They ran the miner out of town and used the coal to pebble the walkways. Old Philadelphia of 1830 was home to a little boy, Stephen Winslow, who grew up to become the leading journalist in the City of Brotherly Love. Stephen Winslow was my great-great-grandfather.

Anthracite coal, called "hard coal," was formed just like any other coal, from the action of sunlight on the leaves of green plants. The process of photosynthesis converted light energy into organic compounds, into carbon, which was then buried in bogs, compressed, and subjected to the heat and pressure of the earth. Anthracite, formed under the additional heat and pressure of mountain building, is a more concentrated form of carbon. Coal is the rock that burns, but hard coal is difficult to ignite. Hard coal came to be called stone coal.

Stones are simple and they are plain and they are everywhere. Common as sparrows or mud or brown leaves. Common as copper pennies or raindrops. Common as the people Grandma used to look down upon as common people. What then, is the Philosopher's Stone? One thousand years ago, alchemists called the Philosopher's Stone the Red Stone. The Red Stone is a common stone that has been infused with an element of spirit. Perhaps it stands for the common man or common woman touched and somehow altered by the world's mystery.

My sister's tombstone stands alone in a Quaker burial ground that, before her, had received no burials since the 1930s. It is a small cemetery squared off by high oak trees and surrounded by flat fields. Nearby stands the white-painted wooden Meeting House. The older burials had no tombstones, perhaps in keeping with Quaker simplicity. As for my sister, the tombstone carver insisted that her name be placed below my parents' names, for this is also to be their final resting place. It is customary, he explained, for the child's name to follow that of the mother and father. But my father insisted that her name be the first one, and so it is: 1946 ~ Susanne Barbara Long ~ 1986.

It is the Jewish custom to leave stones at a grave to make a sign that the grave has been visited. My friend and I, sipping Oregon Chai tea at Starbucks, agree that stones stand for eternity—since any small stone is millions of years old—and that they can be quite beautiful. In Christian funerals flowers are used to symbolize life. But flowers also symbolize the briefness of life, petals curling and browning and drooping to death. My friend's granddaughters Cori and Blake were two beautiful children who went down on Alaska Airlines Flight 261, which crashed into the sea on January 31, 2000. The bodies could not be recovered. Three or four hundred people attended Cori and Blake's memorial service. Bowls of stones were passed, and each grieving person chose two. During the long service, we warmed the stones in our palms. At the end, we chose one stone to keep. We returned the other stone to the mother of the two girls.

The prophet Ezekiel recommended that harlots be stoned to death. The men that visit the harlot were not to be punished. Indeed, it was they, along with their wronged wives and daughters, who would stone the harlot. Ezekiel delivered his edicts and instructions, his prophecies, long before the 1970s when prostitutes organized themselves into COYOTE (Call Off Your Old Tired Ethics) for mutual aid and to defend their basic rights. Addressing the harlot, Ezekiel prophesied that the men to whom she had given pleasure shall "bring up a company against thee, and they shall stone thee with stones, and thrust thee through with their swords. And they shall burn thine houses with fire, and execute judgments upon thee in the sight of many women" (Ezekiel 16:40).

When a woman is raped, hit, slapped, violated, she may disassociate, turn numb, turn off her body sense, her proprioception. She feels nothing at all. Her body temperature drops to the point of hypothermia. She cannot feel her own weight against the offending bed. Psychically, emotionally, mentally, she departs the premises, numbs out, turns to stone. Her thoughts slowly revolving may seem to fill the room, disembodied, like an echoing, distant voice instructing her. Forever after, if she lives, her substance is not flesh and blood, but stone. Especially, the organ trapped in the ribcage, the heart. Her heart has turned to stone.

At one time, so the legend goes, Medusa had beautiful hair. She was a lovely nymph who lived beside the sea. One day Poseidon, the god of the sea, raped her. It is interesting how the rape part of the Medusa story is typically forgotten. In any case, after the rape, her beautiful hair turned to snakes, and any man she gazed upon turned to stone. No doubt Medusa took a certain grim satisfaction in turning men to stone. She was a danger to them, and finally one of them killed her. At her death, a winged horse flew out of her head—Pegasus, symbol of poetic inspiration, favorite of the muses.

The British mountaineer George Mallory called the ice and rock pinnacle of the world, "a prodigious white fang." Mount Everest's Kangshung Face rises straight up out of the Kangshung Glacier and soars two miles into the sky. Everest has, Mallory wrote to his wife, Ruth, "the most steep ridges and appalling precipices that I have ever seen." In 1924 he and his companion, Andrew Irvine, died on the mountain after reaching the summit for the first time in human history. Or after failing to reach it. The point is debated.

Irvine was an inexperienced twenty-two-year-old climber, an engineering genius, an athlete most accomplished in rowing, whom Mallory, in some fateful, fatal impulse, chose to accompany him. On June 7, 1924, Mallory and Irvine failed to descend into the arms of their awaiting team. Seventy-five years later, a search team found Mallory's sun-bleached body. He had fallen down the Northeast ridge, called the Yellow Band, which consists of great downward-sloping slabs of yellow limestone "streaked through with veins of a whitish quartzy rock." His body, with its good leg crossed over a broken leg (he was alive at the end of the fall), lay on the ridge of a dark gray limestone precipice. He had a severe puncture wound in his forehead, the likely cause of death. The body of Irvine, who could not possibly have found his way back without Mallory, has yet to be found.

Tibetans call the mountain Chomolungma—"Goddess Mother of the World." In Tibetan belief, the terrible goddess lives in the mountain. During the Mallory era, two Tibetan porters, obliged to remain overnight in a tent in one of the highest camps yet established, passed the night in dread and swore they could hear the dogs of the goddess barking. Beneath the mountain, according to this tradition, live fire-breathing dragons.

In 1923 Carl Jung began to build the stone Tower at Bollingen on a lakeshore that was long ago occupied by the monastery of St. Gall. He built the tower over many years, stone by stone. The tower seemed to him to represent his inner world, his psyche fully realized. Building it became part of his own process of individuation, of bringing into the conscious, visible world the dark and unconscious parts of the self. As Jung believed the psyche to be ancient and not bound up in a time frame, so the Tower at Bollingen seemed ancient and timeless. He made one turret into a stone room of contemplation to which he alone kept the key, and he spent time there by himself, and permitted no one else to enter. He painted his inner life on the walls of this room. After his wife died in 1955, he built a new part of the tower. Then, he wrote, "I felt an inner obligation to become what I myself am." The tower had an elemental feel to it, eternal as a tombstone, and indeed, at one point during the digging and building, century-old human bones were found.

At the tower, Jung lived in simplicity, tending the fire and the stove, chopping wood, drawing water from the well. Among these stones, he wrote, "there is nothing to disturb the dead, neither electric light nor telephone."

For luck, I carry in my pocket a pebble of stone coal. It is 350 million years old. My amulet was dug from northeastern Pennsylvania, around Wilkes-Barre, not far from where I was born. It is a glossy stone of peacock coal—coal contaminated by sulfur to the blackish green and purple of an oil slick. Coal is light compressed into darkness, its carbon formed by sunlight in the leaves of green plants. The Rig Veda speaks of the time when "the realm of light was still immersed in the realm without light." Coal is that realm of light—sunlight—immersed in the realm without light. Coal is sunlight, buried.

15

Solitude

> No writing on the solitary, meditative dimension of life can say anything
> that has not already been said better by the wind in the pine trees.
>
> THOMAS MERTON

Does the spirituality we crave depend on contact with nature—the woods, the riverbank, the ocean shore? Are we beings of spirit or beings of cell phone and text message? Are we meditators, thinkers, readers, ones given to contemplation, or does the fact that we are now subject to continuous electronic impingement, interruption, and distraction preclude such a life? Does our hyperconnected way of life disconnect us from our own selves? From our own thoughts and values? What is the meaning of solitude in each of our lives? What would happen if we took a bit more of it for ourselves?

Solitude can be delicious. It can be delicious even to a five-year-old. When we—my twin sister and brother ten months older—turned five, our family moved to a farm in Maryland called Ravenswood. At Ravenswood, I got a room of my own, an attic room with steep gables and a square wooden window that opened sideways like a book. My own kingdom at the top of the house. No one was permitted to enter unless invited. It was a plain attic room with an unpainted wide-board floor. It had a narrow bed and a yellow bureau. On the bureau I kept my treasures—stones and bird's nests and acorns. Some summer afternoons, I just lay on my bed in my shorts and T-shirt and did nothing. A breeze rustled the treetops right outside my window. Birds chirped. At the time—this was in 1948—I did not know how to read.

I let my mind drift, with no aim or purpose. I listened to the flies buzz. Once, my Aunt Pat climbed all the way up two flights of stairs to fetch me. I faked sleep so I wouldn't have to descend into the chaos of the household below. In my own room nothing was demanded of me, nothing was required. I was free. I didn't have to set the table or get out the milk or take out the garbage.

We had no money to speak of. And we lived in a din, in clutter and clamor. But my very own room—replete with silence and slanted sunlight—was a temple, a spacious abode where any thought or daydream could turn with the earth's turn toward dusk and darkness. There, in that attic room, I learned what is meant by time enough and space enough.

Much later, when I was in my twenties and living in Boston, solitude had gone the way of childhood. My husband and I were both researching fat books, and our little rooms were cluttered with cartons of research Xeroxes and library books and newspapers. The TV fizzed and droned, often sputtering to itself. I had numerous friends, too many friends, for my deepest desire was to be liked, which made it difficult to turn down any invitation. My husband and I were devoted but quarrelsome, sparring over who was going to cook or wash dishes, debating what movie to see or what rug to buy; arguing over whether to turn off the TV or leave it on.

It was a time of murkiness and confusion—in material terms, clutter and noise. Besides working at menial jobs like cleaning houses (later I became a printer), I had on my days off numerous social "obligations." On Tuesdays I went to lunch with "friends"—people I would not, if I had it to do again, choose to dine with every single Tuesday. Nevertheless, our Tuesday luncheons went on, it seems to me, for years. We went to a white-table-clothed Greek restaurant, spent a great deal of time ordering, and consumed huge platefuls of moussaka or kotopoulo or spanakotiropita plus bread lathered with salted butter plus a glass or two of red wine, as if we were celebrating some great national holiday or as if one of us had just received the Nobel.

In the small times remaining, I worked on my history book. Now this required going into my tiny study and locating the pertinent pile of paper, one of the ever-proliferating (due to my ongoing library research) piles stacked on floor, on the chair, or on the door laid flat on sawhorses that served as a desk. After I found the working pile I would move it to the kitchen table, work there, and when my time was up, return the entire mess to the study.

The time came when I decided to design a better study. I hired a friend who was a carpenter to execute my design. My new study had an L-shaped desk with one surface for handwriting, and another, reached by a twist of the chair, for my Smith Corona Electra typewriter. My study had wide built-in shelves to fit all the research notebooks and papers. The desk and shelves were made of dark Formica—a beautiful hard surface edged with wood painted barn-red. The space was transformed into one of peace and order. I began getting up every morning at 4 a.m. before going to my job running a printing press. In the quiet of that hour, I wrote in my journal and then worked on my history book. A cherry tree bloomed at the second story window. Its white blossoms stood for moonlight and foretold daylight. During these early morning writes I began to remember my childhood. A new composition book, open to clean white pages, invited drifting and dreaming, turning memory this way and that as the earth made its slow turn toward dawn. In this way I began to remember what it meant to have time enough and space enough.

I have long harbored a mild envy of hermits—those whom I imagine living in solitude in a monastery or in a hut by the sea. Yet I'm no hermit, though today I live by myself. At best I get a long afternoon, or rarely, a day, to walk the city looking around, browsing bookstores, writing in a notebook, reading, having a mocha or a latte, letting my mind drift.

Still, I pity people who shun solitude. I pity their talk-show, cellphone, TV, tell-all boredom. As Isabel Colegate writes in *A Pelican in the Wilderness*, a patchwork book of anecdotes about hermits and recluses, "No amount of group therapy, study of interpersonal relationships, self-improvement exercises, personal training in the gym, can assuage the loneliness of those who cannot bear to be alone." They are lonely, I think, because they lack, above all, the company of themselves. To be alone is to be "by yourself," to be with yourself, to keep your own company.

Social life can be so much noise, so much aimless chitchat, interruptions interrupting interruptions. How often do we see two people walking down the street together, one of them talking to a third somebody on the cellphone. How often do we see two people dining at a restaurant, one looking about aimlessly, waiting for the other to get off the phone. At the QFC grocery store the other week, I stood in the checkout line behind a man who never ceased talking on his cellphone while the cashier rang him up and gave him his total, while the bagger asked him his preference for paper or plastic. No eye contact, no connection, just the distracted answer and swipe of the debit card. Both cashier and bagger were visibly offended. They commented quietly on his rudeness while he whirled along, oblivious.

Isn't such a person out of touch, disconnected, more isolated than a hermit?

But what of the real hermits? As a fourteen-year-old, I came across *Alone*, Admiral Richard Byrd's account of living alone at an Antarctic outpost for the duration of the polar winter, April to August 1934. His polar hut was set up a hundred miles south of Little America, winter quarters for the rest of the expedition. Byrd's months spent alone recording weather data in temperatures ranging from 40 to 75 below zero were for him a temporary retreat from a frenetic life of organizing exploring expeditions and raising funds to carry them out. Alone, Byrd spent busy days maintaining his instruments, reading, listening to music on his phonograph, teaching himself to cook flapjacks, extending his snow-tunnels, and working on his (unfortunately flawed) stovepipe. He had twice-weekly radio contact with Little America. For him the months alone in the Antarctic night did not represent a permanent retreat but a chance to "taste peace and quiet and solitude long enough to find out how good they really are."

Byrd's account of the months-long polar night, the brilliant innumerable stars, the spectacular lightshows of the aurora australis, and the feeling of inward harmony he felt with the cosmos are some of the most lucid moments of nature writing in our literature. Tragically, his stove began leaking carbon monoxide, and for two months he struggled to survive in increasing pain, weakness, and despair until his men managed to rescue him.

Even the most dedicated hermit stands in need of other human beings.

Another solitary being to whom I was powerfully drawn was the writer May Sarton. I came into Sarton's *Journal of a Solitude* during those 1970s years of excessive noise and clutter. Reading *Journal of a Solitude* was for me like dipping into a well of peace and well-being. The book recounts Sarton's retreat to a house of her own in a New Hampshire village, a place in which to wake up alone, a place in which to live alone, a place in which to write in blissful solitude. I was especially taken with her descriptions of the rooms, the way the evening light illuminated them, their simple furnishings, the way she kept them swept and washed and polished. And then we come to her vivid portraits of townspeople, the souls who kept her company through the quiet, snowy winter.

No hermit's solitude is absolute.

Indeed, one of the world's better-known hermits, the best-selling writer and monk Thomas Merton, spoke daily to the many visitors to his hermitage and had difficulty keeping them away. But Merton knew solitude. In *The Silent Life* he wrote:

> Not all men are called to be hermits, but all men need enough silence and solitude in their lives to enable the deep inner voice of their own true self to be heard at least occasionally.... For he cannot go on happily for long unless he is in contact with the springs of spiritual life which are hidden in the depths of his own soul.

Merton longed for solitude perhaps more than he actually practiced it. But he also longed for readers.

As did Admiral Byrd. As did May Sarton.

But total isolation, unchosen isolation, can be terrifying and can even damage the brain. Even relatively short periods of solitary confinement, according to the psychoanalyst Anthony Storr in his book

Solitude, very often lead to mental breakdown or emotional instability that can last for years. The very few who can resist and survive do so because they have the ability to turn inward to a rich mental life.

Edith Bone (1888–1975) was a Hungarian-born translator, a British subject who was an outspoken but loyal member of the Communist Party. In 1949, on her way home to London from a translating job in Hungary, she was falsely arrested (as a supposed British spy) at the airport, and spent seven years in solitary confinement under cold, filthy, malnourishing conditions. Her book, *Seven Years Solitary*, is her vivid account of those years. She was sixty years old. For five months she was kept in total darkness, and for the first three years she was denied books, newspapers, and her eyeglasses. She suffered physically (she was ill for much of the time), but virtually thrived mentally. (It helped that her family members lived outside Hungary so that authorities could not further persecute her by persecuting them.)

She refused to "confess" and developed her art of refusal, often taunting her guards or playing dead or making demands or going on hunger strikes. The project of confounding and annoying her guards entertained her, and she otherwise went to her imagination for mental stimulation. She took walks to every city she knew well, thoroughly canvassing the streets, parks, and shops, and always stopping to visit any friends she had there, whether currently living (which she had no way of knowing) or known to be dead. She made an abacus out of inedible black-bread crumbs and made a count, working in alphabetical order, of the words she knew in each of the seven languages in which she was fluent (in English she arrived at an astonishing twenty-five thousand words). She made up what she called doggerel, and repeated these rhyming poems to herself several times a day, so as to not forget them. Her many mental projects kept her so busy that some days passed rather quickly. Edith Bone was released on November 1, 1956, after students involved in the Hungarian uprising captured the prison and processed the detainees for release.

Hermits by choice also turn inward for mental stimulation, but they choose company upon occasion. Searching the Internet on "hermit" I came across a huge website, www.hermitary.com, rich in resources for hermits and aspiring hermits. It is an online community of hermits, no irony intended, with tens of articles on solitude and silence, a blog through which solitary people can connect to other solitary people, and a cache of news stories on hermits in history and around the world.

For me solitude is about turning inward. My sometime-practice is to halt the whirlwind day, go to my bed-sitting room at the back of the house, and sit on the futon. I look at the Carol Ann (of Santa Fe) watercolor I own, which depicts a foggy morning on Puget Sound, boat shacks and madrona trees. I take in the blue stillness of that watercolor dawn. I dip into whatever I'm feeling. I let thoughts drift. I breathe. I listen to the world, whether it drones like an airplane or rustles like a squirrel in the backyard. I could sit there forever, I feel, but after a short time I get up and go back to work.

Early mornings are routinely for me a time of solitude. At 4 a.m. or 5 a.m. it is dark and quiet. No traffic. Open the door to the screened-in side porch and listen to the breeze in the trees. On some mornings the moon is brightly out. Always, the smell of coffee. The birds are still sleeping. I do the yogi "sun salute" *sin* sun, pour the coffee, go to the room I call my scriptorium and begin to write. To do so I must turn on a lamp, but when the sky fades to indigo, then pales to pink, I turn it off and write in the dusk of dawn.

One morning—this was after sunrise—I witnessed a massive gathering of crows to the high branches of the big-leaf maple across the street. Crows and more crows flew in, cawing and flapping their black wings. Crows dwell in roosts that accommodate hundreds of birds, but each morning the roost disperses in family-teams to hunt for food. Crows mate for life and adolescents stay with their parents to help raise the new hatchlings. In a large roost there are many interrelationships— crows in danger aid one another, and they greatly enjoy playing. These intelligent birds utter many different chortles and caws; studies have found different roosts to speak different dialects so that it is difficult for crows from distant roosts to understand one another. This morning should have been the dispersal time; in any case our big-leaf maple is not a roost. But for some reason hundreds of crows arrived. They flapped and squawked for perhaps half an hour and then they were gone. Later that morning, when I left my house, I discovered a crow dead on the parking strip. So that was the keening of crows.

We need our attachments. We need our loves and our intimate friendships. We need those who can hear us. We writers need our readers. We need the people who care for us and we need to care for them. We need company, companionship, good talks over good wine. We need our poetry workshops, our gossip sessions, our rap groups. We need our good buddies and our old friends. We need the merchants and clerks and neighbors and math teachers and road workers we encounter in our communities. We need our communities. We need our colleagues and our fellow workers. We need at least some of our fellow commuters. But we also need solitude. We need times of no talk, no radio, no TV, no phone, no traffic, nobody else. We need time to reflect on the day or the year, to just sit with our breath, to just sit with the light. It is solitude that gives us the world of crows, living their intense, parallel lives. It is solitude that gives us the indigo light of early morning, the slow progress of the seasons, our own slow thoughts. It is solitude that gives us time enough and space enough for contemplation, for reverie, for the company of ourselves.

16

Object & Ritual

If you can imagine that you're a rock
all your troubles fall away.

AGNES MARTIN

Object

The object I am thinking of is a bucket. A bucket is an elemental thing, like integrity. Buckets are sturdy but hardly fancy. A simple bucket is something to contemplate during times of chaos and war. Buckets are found not in boardrooms, but in barns, sheds, and backyards.

I am writing this piece on buckets during the American war in Iraq. I am not fighting in the war, and the war has so far caused me no personal loss, but its images of violence and pain are present in my mind. I counter the disturbance by thinking of a simple object, a bucket, an old bucket.

The first buckets were made of wood or leather, and they were used to draw water from wells. The word bucket entered English from Old English in 1300. A bucket is a vessel with a hooped handle for carrying water or milk. The Anglo-Saxon bucket had metal hoops encircling wooden staves. It had a metal rim clipped with rim clips to the wooden staves. The handle was riveted to handle mounts.

The act of drawing water from a well is a simple duty. The act of milking a cow into a bucket is a simple duty. The word "bucket" has a simple sound. A bucket is a simple object on which to rest the mind. Our times may be disturbed, but within this larger disturbance we can hold a steady course, speak for what we think is right, hold on to whatever seems elemental and essential.

A farm worker carries a full bucket by leaning away from the weight, by sticking out his other arm for balance. This labor, carrying water to water the mare, carrying water from the well to the house, carrying milk from the cow barn to the calf barn is as old as bread on the table. As a girl growing up on a farm I was a great bucket carrier. I prided myself on my strength and on my skill in carrying bucketsful of water, with no spilling. I liked buckets and I liked salt licks and water troughs.

The psychoanalyst Christopher Bollas speaks of choosing objects as a way of choosing the self you want to express in the moment. His idea is that the self is fluid and at least partly chosen. You may choose to pick up a tome by Tolstoy or it may be a baseball mitt—you choose to play ball. These objects—the book or the bat—express different selves at different times. You may choose your garden gloves or a video game or a film. I choose the image of an old bucket. I rest my mind on it.

There are objects that stand for childhood or for particular persons or for lost loves. When I think of my Pennsylvania Dutch grandmother I think of white cotton handkerchiefs with lace edges, printed with violets. When I think of my father I think of his large German shepherd puppy named Toby. When I think of my sister Pamela I think of the streets of Rome.

The object that most represents my childhood is a galvanized metal bucket. I remember our horse Virginia munching oats from a bucket. I remember the pig, its snout in the bucket. I remember the newborn calves sucking on the rubber nipples of calf buckets.

A poem I wrote forty years ago, one of the first I dared to keep, contains a bucket:

Barefoot, squatting in the sunlight
of the backporch
step

flies buzz around a bucket
of old milk
turned to cheese

white barns
cool inside
sun-silenced.

Memory: still as stone.

During times of war and chaos, we could do worse than to think of an old bucket. We could do worse than to think of sunlight on a back porch step. We could do worse than to rest our minds on objects that are elemental and still as stone.

Ritual

The ritual I am thinking of is making the bed in the morning. Making the bed represents taking care of things. It is like sweeping the floor or pouring the coffee. There are acts that state, as Bob Marley put it, "Everything gonna be all right." There are acts that state, "Good Day Sunshine." I no longer attend church, but during my religious girlhood I repeated every day: "This is the day that the Lord hath made. Let us be glad and rejoice in it."

The ritual I am thinking of is to perform ablutions in the dark of morning before the sparrows begin cheeping, before traffic begins its hushing and moaning out on the street in front of the house. To perform these ablutions is to repeat an ancient act of purification. To wash is to hope. The person without hope, the severely depressed person, does not wash. To perform morning ablutions no matter how we feel is to commit an act of faith. The new day is about to begin. This is the day . . .

You make the coffee. You fill the kettle with water, turn on the burner, put the filter into the filter holder, measure out the ground coffee. You pour the bubbling hot water over the grounds and breathe in the aroma of fresh coffee.

You sip the coffee, black and steaming. You write in your journal. You form letters to make words to fill the white page. You listen to the scratch of the fountain pen on the page. You watch the sky turn indigo blue, then robin's-egg blue, then pink and pale white. You write and so reenact a gesture begun long ago by the Sumerians, by the Egyptians, by the Phoenicians, the Mayans, the Chinese . . .

I do not turn on the news in the morning. I do not answer the phone. On September 11, 2001, the morning of the twin towers disaster, my brother-in-law, a news junky, rang my phone over and over until I, thinking something must be wrong, answered. Something was wrong.

Acts of extreme violence—on 9/11 some three thousand lives were lost—do violence to our mind, do violence to our sense of ordinary well-being. I did not know anyone who died in the four airplanes or in the collapsing towers, but their pain, their deaths live in my memory. Such events, for those with no immediate loss, stir memories of horrors personally experienced: the rape, the sister's suicide, the battlefield, the auto accident, the mugging, the murder. On the day of 9/11, trauma-tized vets were re-traumatized, and persons prone to anxiety attacks suffered their anxiety attacks.

Such events disturb the mind, elevate the blood pressure, rattle the nerves. But even during such times, simple acts like sweeping the floor or watering the plants or feeding the dog seem to say, I am alive and I favor staying alive. I favor life.

I am here at my desk writing and we are now at war in Iraq. We are at war in Iraq although it was Al Qaeda, and not Iraq, that attacked the twin towers, and although Al Qaeda was a sworn enemy of the Saddam Hussein regime. We are at war in Iraq due to "weapons of mass destruction," which did not actually exist. During one month of this war (June 2003), the Associated Press enumerated the deaths of 3,240 Iraqi civilians. Meanwhile, by this time (I am writing this in 2007) more than 3,000 American soldiers have been killed.

Despite these sorry facts, we keep sweeping the floor, cooking sup-per, putting the kids to bed. Simple acts, ritual in their regularity. Acts that see to the maintenance of life.

IV Spirit

Fire is the origin of stone.

By working the stone with heat, I am returning it to its source.

ANDY GOLDSWORTHY

17

Writing as Farming

> Stone of the mind within us
> carried from one silence to another ...
> CAROLYN FORCHÉ

I am a child growing up on the Eastern Shore of Maryland. My father wakes me in the hour before dawn. I pull on shorts and a T-shirt. My father tells me where the herd is, and I walk barefoot into the dark. I follow the shadows of cedar trees along the barbed-wire edge of the front field. Robbie the border collie bounds along beside me. We find the herd in the back field, large shadows, soft shadows. One hundred cows raise their heads. I move closer and Robbie begins herding. The cows heave themselves up in their resigned, dignified manner, front feet first. They lumber through the wide gate into the front field, down the front field along the row of red cedars, through the gate that opens into the dirt lane, across the dirt lane into the milking barn. Without speaking, my father and I head cows into stanchions, one by one. We shut the stanchions with a clang. I feed each cow a scoop of grain. My father washes udders, slips milkers onto cow teats. The milking machines hiss and suck in the dusk of dawn.

I get up in the dark to write. I walk barefoot on oak floors, pour coffee steaming into a white porcelain mug. I write to teach my eyes to see. I write to teach my eyes to see tree-shadows becoming trees, cow-shadows becoming Holsteins, Guernseys, heifers, kickers, old steady milkers. I write to hear the pen scratch across the white page, to hear the rataplan of rain on the windowpane. I write to record news from myself, the night dream, the nocturnal question. I write to visit the shadows of myself, obscure memories, unfelt angers, secret desires. I write to record news of the world, city buses nosing their great slow way down the night street outside my house, buses familiar as cows.

18

Hildegard

I believe in art as a spiritual, health-giving process, not just some style.
MALCOLM MORLEY

Hildegard of Bingen began writing poetry, natural history, theology, medicinal tracts, and canticles in 1140, when she was forty-two years of age. I like to think of her as a solitary mystic—hermit plus oblate—and perhaps this hermit notion is what attracts me to this preeminent woman creator of the twelfth century. But Hildegard was no prototype artist-in-garret. In the twelfth century, people did not spend time alone: lords and ladies shared bedchambers with servants and kin, and people did not retire to private quarters even for sex.

Even anchorites did not live in strict solitude. These anchorites were religious figures who, for the glory of God, entered a small stone hut, usually affixed to the wall of a monastery, after receiving from the bishop a blessing as well as last rites in case they should expire unnoticed. They spent their remaining decades there, in that hut. They were sealed in. Monks or in some cases passersby handed in eggs, bread, and beer through a small window, and they handed out pots of feces and urine through the same small window.

Hildegard was born in 1098. She was the tenth child of noble parents, and when she was eight years old her mother and father tithed her to the church. She was put into a stone hut with the anchorite Jutte. The hut was attached to the wall of the Benedictine monastery at Disibodenberg, situated in eastern Germany not far from Bingen at the confluence of the Glan and Nahe rivers. Jutte and the girl Hildegard lived in this hut according to Benedictine Rule, ritual, and schedule for two or three decades. Jutte taught Hildegard to read and write and to play the psalter.

Days there are when I want to live shut in a stone hut. On these days, the phone jangles and shrieks, and the To Do list proliferates like a form of insect. In the silence of a stone hut you could meditate and you could write in guttering candlelight to the sound of a Gregorian chant.

Still, it's difficult to imagine Hildegard's life. And could Hildegard imagine mine? Would she want to? Certainly, I too am a writer. I too published my first book in my forties. I too get migraines. (Hildegard received visions, brightly lighted and accompanied by illness—scholars discuss to what extent her visions were migrainous.)

But I am no theist. I was cured of dogmatic Christianity by overdose in the Methodist Church of my childhood and in the Moravian Seminary of my adolescence. It is entirely possible that Hildegard would repudiate my Self and Works as the work of the Devil. Though perhaps not.

Many simple facts about Hildegard's life are lost. I do not know what kind of stone her hut was made of, whether granite or basalt or slate. I do not know whether the hut had a stone floor or a dirt floor. Undoubtedly, the light was dim. Did they have a pee pot? What were the seating arrangements? The sleeping arrangements?

Jutte and Hildegard lived in mutual solitude, but they were distinctly part of the Benedictine community. They conducted their lives in a rhythm set by monastery bells, observing the office of matin at 2 a.m., louds at first light, prime at sunrise (beginning the day offices), then terces, sext, and none at the third, sixth, and ninth hours, then vespers at dusk, and compline at sunset. These ritual observances were timed by the ringing of bells. The psalms were sung, in Gregorian chant. It was a life of bells, chants, and poems (psalms).

They lived with focus and in seeming simplicity. There were no bills to pay, no social obligations, no appointments to meet except the appointment with God seven times daily as chimed. The wardrobe was provided—a simple muslin dress and simple undergarments. A frugal diet of eggs, fish, bread, fruit, and beer was passed through the small window in the stone hut, and wastes passed out, the hut ingesting and excreting, as if it were a living body.

Jutte taught Hildegard to read, write, and compose music. Hildegard was given a mentor and tutor, Volmar, who was not much older than she was. Eventually the monk Volmar became her secretary, an intriguing role reversal.

The stone hut inhabited by Jutte and Hildegard became a center of spiritual and personal counseling. People came for advice and for healing. Young girls joined the cloister. The hut became a nascent convent. (These events call into question its physical dimensions.) By the time Jutte died in 1136, Hildegard, at thirty-eight years of age, was clearly her successor, the leader of a convent.

Four years later, prompted by strong visions, she began to write. She understood her visions and their written depictions and explications as divinely inspired. So also did the Benedictine bishops, and eventually the Pope, deem Hildegard's writings direct transcriptions of the voice of God.

She became a leader, preacher, and spiritual advisor, and in 1150 moved her convent to Rupertsberg, near Bingen, on the Rhine River. In this the monks of Disibodenberg opposed her, but after she went into some sort of rigid trance, they determined that her wish was God's will and let her go. She composed seventy plain songs, collected as the *Symphonia*. She composed three theological works (*Scivias, The Book of Life's Merits*, and *The Book of Divine Works*), a natural history, a book of remedies titled *Causes and Cures*, and several biographies of saints. Besides this, she carried on an extensive correspondence.

What, we might ask, were the conditions and traits that nurtured Hildegard's creative work, monumental and moving as it is, even today, even to a non-theist? How did this child, constricted to narrow stone walls and a flickering candle, grow up to become one of the significant creators of her age? What is it that nurtures creativity? What is it that squelches it?

Hildegard's secretary, the monk Volmar, worked with Hildegard from their adolescence until Volmar died in 1173, when they were both in their seventies.

At the death of Volmar, Hildegard found herself "in sorrow and desolation." Her companion (as it seems compelling to call him) died when they were in the midst of Hildegard's composition *The Book of Divine Works*. In the book's acknowledgments she thanks other monks who assisted her after Volmar's death in setting down her visions. These monks "faithfully heard and loved all the words of these visions without tiring, since they were sweeter than honey and honeycomb; and so through the Grace of God, and with the help of these venerable men, the writing of this book was finished."

Hildegard grew up immersed in Gregorian plain song chanted hourly at the Benedictine services. She grew up in a center of increasingly bookish preoccupations: the monks—and the nuns—copied and illuminated manuscripts with great dedication. Hildegard was taught—required—to play the psaltery, the medieval stringed instrument that accompanied the plain song. She and her community believed that her visions and her writing were Very Important. She was supplied with time, food, shelter, writing supplies, assistants. She had at least one intimate companion, and a fully supportive community that witnessed and honored her creations. In return, Hildegard gave her art all that art demands: thorough training, sustained attention, persistence, faith.

She broke through. Growing up in a stone hut, living in stone rooms, chanting, praying, composing, she broke into light.

19

Me and Mondrian

Think of me as simply a poor man who values life and tries to embellish it.

PIET MONDRIAN

What is so remarkable about red and yellow cornered within a carbon-black grid? Or a blue square painted off-center within a geometry of gray and black? And what is it that draws me to the painter of such compositions—Piet Mondrian? When Mondrian died of pneumonia on February 1, 1944, in Manhattan, I was a ten-month-old child. He was seventy-two years old. The Second World War was on. To escape the war, Mondrian had moved to Manhattan and he loved Manhattan. He'd been living in a two-room studio. His estate, his friends recorded, consisted of pipe tobacco and pipes, an impeccable but threadbare suit or maybe two, dancing shoes, two or three pairs of spectacles, five books, a little pile of his own writings, a Dutch passport, orange-crate furniture painted white, a narrow bed, a gramophone, a few jazz records, painting supplies, and about thirteen hundred paintings. At his death his work was recognized by a few hundred persons. A few newspapers had carried reviews of his few exhibitions. Today he is considered one of the great painters of the twentieth century.

Beginnings frame a life. Mondrian was born in 1872. He was the second child born into a strict Calvinist Dutch family, a conventional, small-town Christian family with a strict, domineering father, the headmaster of a strict religious primary school in Amersfoort, the Netherlands. Mondrian was given his papa's name—Pieter Cornelius Mondriaan—and he grew up under his papa's thumb. There was a parental art streak—his father sketched habitually, and his uncle, Fritz Mondriaan, was a professional painter within the Romantic/Realist Hague School. I have at this moment ten books on Mondrian stacked on my living-room rug. Among them, they manage about two sentences on his mother, Johanna Christian Kok.

He wanted to be an artist but his family required him to get a teaching certificate first. When, at age twenty, Mondrian finally arrived in Amsterdam to go to art school, he felt obliged to ask Papa's permission to take a life class. He was at the time a conventional, strict, dutiful Calvinist artist. Over the next decade he became a good but conventional painter of barns and cows and polders and Dutch windmills.

Mondrian's barns and windmills—not the black grids and reds for which he is now known—remind me of my own beginnings. Barns and cows. We even had a windmill, a steel-trestle structure with small metal sails, not Dutch, not scenic, but a working mechanism that pumped water from the well when the wind blew. (When the wind didn't blow, an electric motor worked the pump.) We lived a riverine life, bucolic, among cows. Our parents read to their children every night from the King James version.

As I write this I am sixty-six—the age of Mondrian's death, which occurred sixty-six years ago. My life's trajectory has taken me far from Bible and barn work, but life retains aspects of the old way. Art—whether writing or painting—requires a work ethic. This applies whether you are an ordinary striver like me or a genius like Mondrian. Mondrian had dogged work habits, and I imagine them evolving out of his early Calvinism. Art requires making a path and keeping to it regardless of the temptations to stray, regardless of the hopelessness of the matter, regardless of the unbelievers all around. Mondrian lived a frugal, hermetic life fraught with war. His spars with romance were scant and short lived, but he always enjoyed one or two friendships within the art world. He worked on his paintings the way a medieval monk might work on an illuminated manuscript—with painstaking concentrated devotion.

An artist, chameleonic, takes on the coloration of his age. Mondrian was born into the age of photography, evolution, relativity, radio. In 1902 he was thirty years old. Photography had mooted painting's function to record the scene as it appeared, and this opened the door to abstraction. Darwin's theory of evolution, then fifty years old, contradicted certain details of the Adam and Eve story. It would not be long before Einstein's theory of relativity turned time into a dimension. The radio would not come into its own until the 1920s, but when it did, it brought jazz into the painter's studio.

We lived in the same era, he covering its first half, I covering its second. Although our temporal overlap was only ten months, our philosophical and technological overlap exceeded that since Mondrian created within the context of the avant-garde art world, and I grew up in a comparatively backward country context. In his day photography had relieved painting of its duty to portray; in my day our family lacked a camera with which to portray itself. In his day there was no television (the first commercial broadcast occurred in 1941, but it was the mid-1950s before people actually owned a set). In my day my father opposed television, and there are members of my family for whom the television age has yet to arrive. In his day radio was king; throughout my late childhood I did my ironing chore while listening to country music on the radio.

War divided Mondrian's life into chapters. In 1911 he moved from the Netherlands to Paris. In 1914, just after he returned home to visit his terminally ill father, the Great War began its terrible machinations, and he was unable to return to Paris for five years. He mailed his rent money across the border to pay for his studio, where his paintings stood waiting for their painter. (Back in Amsterdam, he helped found De Stijl, renounced the curve, and began developing his philosophy of abstraction, which he called neoplasticism.)

In 1919 he returned to Paris and lived there for twenty odd years, painting and theorizing about painting. In 1938, believing that Paris was about to be bombed into rubble, he moved to London, where his friends, artists Ben Nicholson and Barbara Hepworth, helped him set up a studio. Just then the London blitz began. Mondrian became so distraught over the sirens and bombs, not to mention the 1940 invasion of France, that he could scarcely paint. When a bomb flattened the building right next to his studio, he decided to move to New York. His young friend (and eventual heir) Harry Holtzman helped him get there.

Manhattan already provided refuge for a number of European artists known to Mondrian. In New York Mondrian was happy. He was admired, he had a few friends, and he received two one-man shows—the sole one-artist shows of his lifetime.

And what of his spiritual biography, his philosophical biography, the trajectory of his painter's soul? In 1909, at age thirty-seven, two years before Paris, he discovered Theosophy and cast off forever his Calvinist-colored Christianity. Theosophy was pantheistic—many gods, old gods, the gods of the East as well as the West. It was spiritualist, believing in meditation and the divine within the self. It was Platonic, believing that objects and beings were manifestations of a spiritual ideal, believing in divine absolutes, believing in universal truths and universal states of being. Theosophy framed Modernism, for here we have the essence of Modernism and the essence of abstraction: to leave behind the subjective, the particular, the figural, the pictorial. To paint the essence, to paint the truth. Among the five books in Mondrian's possession at his death were works by Rudolf Steiner and Krishnamurti—avatars of this spiritualist tradition. I am struck by how strongly Mondrian's belief system affected his aesthetic, not to mention his fanatical dedication to working out the meanings of abstraction through his painting. I am struck, too, by how opposite postmodern philosophies are—concerned with the subjective and with identity, concerned with the subtexts (often masculine and Caucasian) of supposed universal aesthetics. Modernism eschewed biography (high art concerned the pure object, divorced from the life of its artist) and Modernism eschewed nationalism. Mondrian eschewed his father's name, eschewed the ur-Dutch double *a* in his surname. Pieter Mondriaan became Piet Mondrian, who hated war, who hated nationalism, who did not care to be Dutch.

Add to Theosophy—Cubism. The flattening of the picture plane, the repudiation of perspective, the repudiation of naturalism. Mondrian encountered Cubism in 1912–13, two years after his conversion to Theosophy. He was then forty years old. He had another thirty-two years in which to change painting forever.

As a teenager in the 1950s I became entranced with modern art—Van Gogh above all, Picasso, Matisse, Gauguin, Joan Miró, Chagall. The usual suspects. And during my brief, intense, college-student love affair with Manhattan, I saw Mondrians hanging on the walls at MoMA. I passed them by. Mondrian eluded me the way irony and minimalism and money eluded me. I didn't *get* his geometries in black and grey and red and blue, each one titled *Composition*. As for music—and music is quintessential to any understanding of Mondrian—I liked hillbilly, then classical, then Elvis, then the Beatles, then the Rolling Stones, Janis Joplin, the Grateful Dead. Later I reverted to hillbilly—bluegrass, country, old time. In those years I did not get jazz just as I did not get Mondrian. And jazz is the key to Mondrian.

Mondrian had a life-long passion for dancing. He didn't date, he didn't court women, and neither was he gay. But he danced. He took lessons and studied assiduously the tango, the shimmy, the Charleston. (When the Netherlands threatened to ban the Charleston, Mondrian threatened to never return.) His whole life, he frequented cheap dance halls, and he danced. He believed that jazz was the music of the future. He owned a gramophone, which he painted red, and he owned jazz records.

What else? Mondrian grew to detest curves and, perhaps related to this, one biographer states outright that he was a misogynist. Speaking of the art trend known as Futurism, Mondrian wrote, "In a futurist manifesto the proclamation of hatred of woman (the feminine) is entirely justified. It is the Woman in Man that is the direct cause of the dominance of the tragic in art." By tragic he meant, one of his biographers explains, "impulse, romanticism, sentimentality, mannerism, baroque, everything, in short, that he detested."

Woman has ruined man, women have ruined men, and of course, the Woman in Man has ruined art.

Well. No one is perfect.

Mondrian grew to detest the color green, to detest trees (which he had once lovingly painted), to detest the out-of-doors, especially parks, the countryside, any suggestion of rural scenery. He grew to love cities and particularly the street grid and the high-rise right angles of Manhattan. His spiritualism was abstract, colored by the "non-colors"—gray, white, and black—and by the primary colors. He arranged his sparsely furnished live-in studio as if it were a walk-in painting, impeccably clean, painted white with red, gray, blue, and black accents.

His biographers use words like "crotchety," and "elderly bachelor." My favorite story about Mondrian is the one about the Paris tax collector attracted by the painter's budding reputation to investigate whether he might owe taxes. This meticulous civil servant was moved to pay a surprise visit to Mondrian's studio, where he found the painter "busy preparing his inevitable soup. Faced with the monastic starkness of the studio and the revolting brew simmering on the gas-ring, the disconcerted tax-man withdrew without further investigation."

I am drawn to Mondrian's fanaticism. I am drawn, even, to his penury, his will to put art first. I am drawn to his propensity to be reclusive and to his love of dance. I like the way he made his living-working space into a painting. I am drawn to his vision of how life should be lived, seeking the essence, working by hand on a single painting for months. He worried money and needed money, and he did sell a painting from time to time, but for him art was a way of life, not a "business."

Why return to Mondrian? What do his abstractions mean to us, more than sixty years after his death? His geometric designs look conventional now, mainly due to their tsunamic influence on graphic design, architectural design, and fashion design. Mondrians do not reproduce well in print. To see a Mondrian, it is necessary, I think, to stand before an actual painting. The reproductions look flat and bland and graphic. In print the dynamism, the kinesthetic tension, the rhythms and pushes and pulls are lost.

When Mondrian moved to New York, he discovered boogie-woogie. He immediately set about learning to dance the boogie-woogie. I of the bucket feet find myself innocent of boogie-woogie. What is boogie-woogie? What was it in the early 1940s? I go to the Internet, to that instant music-history library that is YouTube. I type "boogie woogie" into the search box. Oh my. Boogie-woogie piano is electric, shocking, exhilarating. And the intricate, athletic, fast-footed dance that is the boogie-woogie takes your breath away. To imagine Mondrian the man, imagine him dancing the boogie-woogie. One of his dance partners, "Mrs. Stuart Davis" (wife of artist Stuart Davis, probably Roselle

Springer), said, "He danced very well, but he did steps too complicated for me to follow."

Referring to his paintings, Mondrian spoke of "dynamic equilibrium." The paintings have energy, tension, rhythm. They are organic, not mechanical. (In fact, Mondrian worked not by mathematics or measurement but by intuition.) Mondrian, wrote biographer Frank Elgar, "behaved as if he hoped to wrest from the object, through patient exploration, some secret substance, an intimation of the absolute, which he sensed in all things, and especially in the humblest, the most common. Much later, he discovered that this secret, this divine absolute, lies in us, and that it is we ourselves who for the most part project it into things. He would then be led to discard progressively the natural aspect of things, and strive to paint the divine itself, the absolute as such, even as he found it within him, without reference to any external object."

When, in his last, New York years, Mondrian dropped his signature black grid, when he made bright new grids by stacking up small rectangles of red, blue, and yellow, when he titled his last paintings, not *Composition* but *New York* and *Broadway Boogie Woogie* and *Victory Boogie Woogie*, was he returning to the referential? I don't think so. His paintings jive, they move in a syncopated beat, they realize a dynamic equilibrium. They do the boogie-woogie, and that is their essence. More than that, they *are* the boogie-woogie.

20

On Quietness

Sometimes the world of things has something to say. Randall Jarrell wrote
that stream water made a sound that was like a spoon or glass breathing.
MARY KINZIE

My backyard is quiet and that's one reason I like it. In fantasy I'm a
hermit. I live out back in a hut, and my poems are my prayers.

But in life my "home office" buzzes and dings with computer and
printer. I'm a person—writer, teacher, editor—who's too busy, over-
scheduled, often interrupted, and seldom caught up. Perhaps that
explains why I crave quietness. Or could be it's a common human crav-
ing. I like the idea of quietly writing at a heavy oak desk, the oak thick
enough, solid enough to emit quietness.

Quietness is simpler than silence. Or perhaps simple is not the
word I want. It's more familiar, more homey. A quiet night at home
might include washing the dishes and reading by the fire. It might
include quiet music, quiet conversation, quietly sitting. A quiet day
might be a day of cooking and gardening. It might include sweeping
the sidewalk.

Silence is more forbidding, perhaps a bit fearsome. A silent night
is a holy night. There is such a thing as getting the silent treatment.
You can be greeted with a stony silence. To be silent means to refrain
from speech. To be silenced is to be repressed, suppressed, censored,
shut up.

To be quieted is to be calmed down. The Anglo-Norman and Middle
French root of quiet (*quiete*) contains quietude—tranquility.

There are artists who capture quietness in their works, and gazing at their works quiets the mind. One reason I like going to art museums is to quiet my mind. I like going alone and I may not stay for long.

Here at Seattle's Henry Art Gallery I stand before a large-format photograph (4 by 5 feet) of a dry West Texas landscape. A barnyard, fenced with a rough-stick coyote fence, gated with a wide-swinging barnyard gate. An expanse of gravel and dry grass. The vast Texas sky. Close-up, a truck fragment—tire, chrome fender, a blur of red. A shed, shot from ground level, with the rippled roof-edge evidence of corrugated tin. On this dry ground sits a tiny (life-sized) brown-capped bird. The sun is hot. It is quiet, very quiet. You have entered this quiet country and you see it through the bird's eye. The photographer, Jean-Luc Mylayne, will spend two or three months to get such a picture. All twenty-three of these large-format "landscapes with human traces" include a small bird. Mylayne chooses a spot where birds flock, chooses a particular bird for his subject, and allows the flock to get used to his presence and equipage. He names his large-format photographs according to the time spent—"No. 198 January February 2004."

Is it Mylayne's long quiet days with the birds that communicate quietness to the image and through the image to me the viewer?

Another day in another museum I go looking for another quiet image. I am on a search for what a "quiet image" might mean. Alas, this is family day at the Seattle Art Museum, the day to "Rome the World." The museum is noisy, chattering and laughing, baby-crying, replete with running feet and parental reprimand. I look for quiet corners and quiet images but nothing is quiet. Is it possible to find quietness amid noise? I believe it is, but not for me, not today. This quiet object I seek—does it exist? Is this quest for quiet entirely subjective, entirely in my head, my own emotional problem or psychological fixation?

No. I'm sure of it. Certain objects emit quietness: wooden spoons, diner mugs, a bowl of pears, old bones, thick books, rocking chairs creaking on old porches.

Wandering the museum I feel agitated, dissatisfied, slightly lost. This entire museum contains not one quiet thing to look at. I wander about in a desultory manner and then go to lunch at the restaurant.

The restaurant is loud like a school cafeteria. But I get seated and have a chicken salad sandwich and a bowl of potato soup and I begin to feel better.

Then I go up to the fourth floor away from most everybody and there I find a quiet object. It is an ancient figure carved in white marble, female, about a foot and a half high. Her head is oval, featureless except for a triangle nose, abstract-looking, modern-looking. She has small breasts, arms folded over her torso, long thighs, knees oddly bent at a slight angle, toes pointing straight down (was she once viewed from far below?). She dates from 2500 BCE, the Greek Cyclades period. Does her age—human hands carved her more than 4,500 years ago—contribute to the feel of quietness she emits? There's no ornament, no fuss, no striving, no name, little information. Was she a goddess or was she a girl? Later I google "Cycladic Art" and discover that these were funerary figures, placed on their backs, face up. The stripped-to-essentials abstract-art look results in part from the fact that the paint has worn off.

Gazing at this ancient figure quiets me. Somehow, deep quietness is related to living a meaningful life. Or is "meaningful life" too heavy a term, too sweaty and forced, almost banal, leading to the uninteresting challenge: Meaningful how? Meaningful to whom? Say, rather, that quietness opens the door to a richer interior life. I cross the threshold into that quietest of rooms and here are the muses in their simple garments. Will they be kind? Will they attend to my case? Will they help me compose a quiet sentence?

Is there such a thing as a quiet sentence? Writing emits sound like a tuba or like the wind, but only when it's read, whether silently or out loud. Bam! Bam! Eek! Don't shoot! Those are loud sentences.

So how would a quiet sentence sound?

Quiet sentences doze in somber shadows. They have sipped smooth Irish whiskey. They mosey along toward nowhere. Across their flat white plain, they softly sigh. They move so slowly because they make their own meanings and keep their own time. Their words drift and curl like mist among the locust trees.

Our *Homo sapiens* brain needs the trees, needs to meditate while sitting under trees. Our neurons generate rhythmic electrical pulses (brain waves) including alpha waves (8 to 12 cycles per second) emitted during untroubled focused attention. Alpha waves pulse when we are aroused above sleep but below anxiety, stress, and other states of being stirred up such as thinking and learning. And it's not only alpha waves but the synchronicity of neurons thrumming in unison from different parts of the brain that characterizes mindful attention. Synchronous brain oscillations suggest quietness.

One more museum: Seattle's Asian Art Museum, Buddhas, Bodhisattvas (those saintly souls who forego Nirvana in order to stick around and help others attain enlightenment). I sit before a life-sized wooden female Guan Yin, the Bodhisattva of Compassion, carved in tenth-century China. She is seated, meditating, one knee drawn up, one foot on the ground. Her mudras (hand positions) on the one hand touch the ground (steadfastness) and on the other hand, bless. I sit before her on the bench provided. I have my cynical thoughts: What temple was this sacred figure stolen from, when, and by whom? What is a tenth-century Chinese deity doing in twenty-first-century Seattle?

I look up at her. I decide to stay here for twenty minutes, to take in her blessing. The artist who carved her must have felt her quietness. Many hours, many years of learning to carve so skillfully her robe, her benign face, her benevolence. I breathe. I feel quieter and quieter. I must ignore another museum patron entering the gallery. I do ignore. I sit alone with Guan Yin. My breathing slows. I can almost feel my brain slowing to alpha. Now I'm really here, hushed, attentive—receptive to Guan Yin, her quietness, her eternal quietude.

21

Notes Composed in the Dark of Our Time

Dreams of a better life are inseparable from the good life ...
KARSTEN HARRIES

Where am I going? Where are we going?

Every day, every twenty-four hours, an estimated 150 species—of plants, insects, birds, mammals—become extinct. This according to the United Nations Environmental Programme. Of course the numbers are debated, just as if these species were civilians lost in a war. Still, seven out of ten biologists consider that our current massive loss of species (the Sixth Great Extinction) poses a major threat to human existence in the next century. This according to the American Museum of Natural History. We know what the problem is: global warming. Loss of ecosystems to agriculture and to urbanization. Invasive species. There you have it. We know it.

What about hope? Is there anything to hope for?

Is it crazy or naïve to hope for a world replete with tadpoles and turtles and bluebirds and butterflies and bald eagles and salmon and sea otters? A world replete with forestlands and grasslands, prairies, deserts, salt marshes, mountain meadows. A world replete with diverse ecosystems—ecological niches not coated in concrete, not covered in monocrop. Cities lush with trees and gardens as well as people. Clean water. A world where people can live with respect, good food, sanitation, education, shelter, health, liberty. Dare I mention beauty?

Pretty hopeless, wouldn't you say?

In March 2013 the *New York Times* reported that the number of monarch butterflies (*Danaus plexippus*) to complete their annual migration to Mexico was the lowest in two decades. That orange and black beauty. Butterfly royalty, somebody called it. In the Midwest, even as far north as Canada, they lay their eggs on milkweed. As winter approaches they fly hundreds of miles southward to spend the cold months clinging to branches in the forests of Mexico. Drought in the Midwest has decimated the monarchs, and so has the eradication of milkweed. Midwestern farmers have planted corn and soybeans genetically modified to tolerate herbicides so that farmers instead of weeding can poison weeds without harm to the crop. In the process they are poisoning milkweed—monarch food, the only plant monarch butterflies lay their eggs on, the only plant monarch caterpillars can feed on. About 7.5 million acres of milkweed has been lost. Added to this, the monarch's wintering-over forest in Mexico is shrinking due to illegal logging.

Is there any hope for the monarch? Yes there is. In the winter of 2016 monarchs covered about ten acres in their winter forests, more than three times as many acres as during the previous year. Yet the butterflies remain vulnerable: at their peak twenty years ago they covered forty-five acres of forest. If the monarch survives, it will be thanks to the Kansas-based Monarch Watch and other conservation groups, gardeners, school children, and regular people who are planting milkweed in order to save the monarch butterfly. And it will be thanks to the United States Fish and Wildlife Service, which is working to plant milkweed (250,000 acres in 2015) and to reduce the use of herbicides that destroy it. And it will be thanks to people in Mexico (along with the Mexico government's environmental watchdog, Profepa), who are working to halt the illegal logging in the forest reserve where monarchs hibernate. And so we live between hope and despair.

The largest single monocrop in the United States is the grass lawn, divided among private yards and public parks and other public spaces. This is a suburban and urban disaster for ecological diversity, and it amounts to as many as 40 million acres, according to the entomologist Douglas W. Tallamy (*Bringing Nature Home*). It's bad for insects, and, as Tallamy tells us, we need insects, and not only butterflies and honeybees. We need insects because birds need insects. For birds, nectar and berries are not enough. It's insects they need to nourish their nestlings with protein. And insects require native plants. They evolved with native plants, and many survive poorly or not at all on the exotic (nonnative) grass lawn or on the hundreds of other exotic species we grow in our parks and gardens.

Douglas W. Tallamy. *Habitat destruction as a result of anthropogenic changes is a huge problem everywhere for life on earth. That is precisely why we can no longer rely on natural areas alone to provide food and shelter for biodiversity. Instead, we must restore native plants to the areas that we have taken for our own use so that other species can live along with us in these spaces. We can start by restoring native plants to our gardens.*

The planet is heating up faster than expected. According to *Cooler Smarter*, an excellent book published by the Union of Concerned Scientists, the decade 2001 to 2010 was the hottest decade on record. And in January 2016 scientists announced that 2015 was the warmest year since global recordkeeping began in 1880. Fifteen of the sixteen warmest years on record have occurred since 2001. And it gets worse. In March 2015 the global concentration of carbon dioxide in the atmosphere surpassed four hundred parts per million. This is more carbon dioxide in the air than has been the case for several million years. And here, we ourselves as a species *have only been in existence for* two hundred thousand years. Fact disbelieved by many: excess carbon put into the air by our use of fossil fuels is the prime engine of global warming. The particular isotope of carbon in the carbon dioxide building up and over-warming the earth is distinctly that produced by our use of fossil fuels, not from sun flares or volcanoes or whatever. Also, it's the troposphere—the lowest level of the atmosphere—that's getting hotter, not the stratosphere, as would be the case if sun flares were over-warming the planet. All of which needs to be repeatedly repeated due to public confusion over climate change, resulting in part from an enormous disinformation campaign funded by ExxonMobil. What the $16 million campaign spreads is not contrary facts but "uncertainty" about the science. Go to a report available online titled "Smoke, Mirrors & Hot Air: How ExxonMobil uses Big Tobacco's Tactics to Manufacture Uncertainty on Climate Science," published by the Union of Concerned Scientists. Go there and weep.

Without hope, it's difficult to take action or even to function. It's difficult to get out of bed. When things *seem* hopeless, they *are* hopeless. Action begins with hope. Here are some reasons for hope.

At Salina, Kansas, the people got together and entered into a competition with neighboring cities to see who could lower their energy bills the most. Salina lowered its carbon dioxide emissions by 5 percent. Not every Salina citizen was convinced of global warming, but most were convinced that we want to reduce our dependence on foreign oil and lower our utility bills.

The Cayuga River in Ohio was the most polluted river in the United States. It was so polluted that in 1969, it caught fire. All the fish in the Cayuga were dead. The fire sparked a movement to clean up the river. Today the Cayuga River "boasts 44 species of fish" and provides a haven for picnicking on the shore and swimming and boating in the clean water.

There's the move to retrofit buildings to make them energy efficient and green. Anthony Milkin retrofitted his building—the Empire State Building in New York City—and made it one of the top ten buildings in the nation for energy efficiency. Its yearly carbon dioxide emissions are down by a hundred thousand tons per year. The energy bill for the building is down by $4.4 million per year.

These reasons for hope are offered in the book *Cooler Smarter.* Here are two more reasons for hope: not one, but two of my friends have installed solar panels on their houses. On both houses, the meters have started running backwards. Their home-funneled sun is fueling the grid.

There are seven billion people in the world. And just what exactly are we seven billion people to eat? Don't we need the monocrops and don't we need the Concentrated Animal Feeding Operations (CAFOS), each milking three to seven thousand cows—cows with their tails cut off for easier milking access, cows living on concrete 24/7, cows routinely fed antibiotics to ward off bacterial infections common in overcrowded CAFOS, cows producing too much manure to spread so that CAFO manure lagoons in the Yakima Valley, in Washington state, for example, have grown as big as two football fields? Don't we need genetically altered, insect-resistant, herbicide-resistant crops? How are we to feed the world's people?

We Americans throw out 40 percent of our food. Most goes into the garbage, which goes into landfills, which emit methane, which warms the globe. A landfill is not a compost bin.

Here's a reason for hope. Ron Finley is a gardener from south Los Angeles, a place of trash, boarded-up buildings, vacant weed lots, graffiti-coated walls, fast food, obesity, and poverty. Finley and a group he helped to establish, LA Green Grounds, gardens in parking strips and vacant lots, many owned by the city. He began a movement to plant vegetables—beans, broccoli, peas, squash, watermelon—and tall sunflowers and fruit trees. The community joins in, the food is free, the neighborhood is becoming downright beautiful, at least part of it is. To hear Ron Finley talk, go to TED Talks. Hope lies in neighborhoods, communities, and individuals doing something, taking some action, no matter how small. Hope lies in local farms, garden plots, pea patches, yards turned into native-plant gardens, maybe with a grass path.

My Seattle backyard runs narrow as an alley behind the house. A rusted chicken-wire fence demarks my plot from my neighbor's bigger plot. On my side of the chicken-wire fence, for the first twenty-two years I lived here, there grew a row of twenty multitrunked English laurel trees. This shiny-leaved solid wall darkened the yard and obliterated the view into the adjacent yard. At one point the laurels were so tall and they leaned so far toward the house that roof rats repurposed them as rat roads to the roof. (I have bid farewell to these rats, never mind the details.) A few years ago I hired a man with a chain saw to cut this so-called hedge down to about five feet tall. Back it grew.

Now I discover that English laurel, sometimes called cherry laurel (*Prunus laurocerasus*) is on the monitor list of the Washington State Noxious Weed List. It's a "weed of concern" in King County. It's native to southeastern Europe and southwestern Asia, meaning that here it has no natural predators. Birds void its berry seeds and spread it and we spread it by dumping English laurel as yard waste, from which it re-roots and spreads into woods and parks, obliterating one more eco-niche. It's the second most common invasive tree species in King County, Washington, after English holly.

According to biologist Niles Eldridge, invasive species have contributed to 42 percent of all threatened and endangered species in the United States.

Where I live, a most invidious invasive is English ivy (*Hedera helix ssp. hibernica*). It's "Seattle's worst weed . . . It strangles trees, smothers delicate native wildflowers, blocks sunlight, sucks up tons of precious soil moisture, harbors rats. . . . Many local ravines that originally supported a rich mosaic of varied wildflowers . . . are now ivy ghettos" (Arthur Lee Jacobson, *Wild Plants of Greater Seattle*). Yet—you see it everywhere planted in yards and gardens. It covers walls, climbs trees. Nurseries offer it for sale. Low maintenance!

I purchase a razor-sharp, razor-toothed pruning saw with two nasty-looking rows of teeth. You might want to know what I have against chain saws. Nothing. I'm just terrified of them. I begin cutting down my English laurel with my hand pruning saw. Seattle Public Utilities collects yard waste each week, and requires branches, trunks, and stumps to be four feet or fewer long and four inches or fewer wide. Many of my laurel trunks must be cut twice. It takes me about three months, working for an hour on most days. I hereby report that the row of laurel trees—20 trees with a total of 120 trunks—are now gone. I further report that my deltoids and triceps have gone hard. My new backyard is sun struck but still stacked with laurel brush, my laurel-cutting fetish having run ahead of Seattle's collection schedule. Every week I put out a bin of laurel trash, and every week Seattle Public Utilities pulverizes it and turns it into compost.

This is a reason for hope.

I dream of a better world. I dream of a better yard. What if my dirt backyard were ferns and flowers fluttering with butterflies? Butterflies can drink the nectar of any flower, but their caterpillars can typically feed only on the plant they evolved with. If I plant dogwood, the butterfly spring azure may arrive. If I plant lupine, it will be the silvery blue. If buckwheat, the acmon blue will arrive. Gooseberry will bring a flock of tailed coppers; dock, the great copper. I could plant violets to lure the western meadow fritillary or asters to draw the northern checkerspot. I dream of a garden fluttering with butterflies, but not the monarch, which, though orange and black and gorgeous, is not native to the Pacific Northwest. Maybe a painted lady . . .

English laurel roots run big and deep and they're difficult to extract without extensive digging, probably with a backhoe. I was going to poison my English laurel stumps with Roundup, but my new neighbor, a single mom named Cristina, insisted that it's not true what they say, that Roundup does not affect the soil around plants it's used to kill. I don't know the truth about Roundup, but I promise Cristina I won't use Roundup or any other herbicide. As new leaves sprout out of the cut-to-the-ground stumps, I crush them. I don't know how long it will take to kill these stumps, but I do know that trees require leaves to live. This laurel extermination project, by the way, is fun.

What's next? I plant a red-flowering current. I plant a mock orange, I plant a Cascara tree. I add a deer fern and a maidenhair fern to the ferns already present (two sword ferns, a wood fern, and a lady fern). I decide on a serviceberry tree. I plant a huckleberry bush. I plant salal in the shade. Out on the parking strip, I plant Oregon grape and kinnikinnick.

All native. This will not be an entirely native garden, which would be a nice fiction in any case, but all the new plants will be natives. And next, the butterfly garden: Douglas aster, pearly everlasting, lupine . . .

Howard Zinn. *The word "optimism" . . . makes me a little uneasy, because it suggests a blithe, slightly sappy whistler in the dark of our time. But I use it anyway, not because I am totally confident that the world will get better, but because I am certain that* only *such confidence can prevent people from giving up the game before all the cards have been played. . . . Not to play is to foreclose any chance of winning. To play, to act, is to create at least the possibility of changing the world* ("The Optimism of Uncertainty" in *Failure to Quit*).

Will the yardwork I'm doing make the slightest bit of difference?

I have no idea. I do know that it has happened in the past, in history, that enormous changes have burst forth from small beginnings, unpredictably. Consider the Civil Rights movement. Consider, for that matter, the environmental movement. And there are many people—how many I do not know—making small and not-so-small beginnings, clearing invasives, creating niches for bugs and birds, restoring to natives shorelines and woodlots and backyards. Planting food and flower gardens in urban neighborhoods. Eschewing insecticides.

Besides, what better place to begin than in your own backyard?

22

Balancing Act

In my book poetry is a necessity of life, what they used
to call nontaxable matter.

C. D. WRIGHT

On March 17, 2003, the day I turned sixty, the call came. My mother
had taken a turn for the worse. Her dying preoccupied the next weeks,
and after her death on May 29, 2003, I thought of her every day. These
events—my mother's life and death—superseded my big birthday. It
wasn't until the following year, when I turned sixty-one, that it hit me.
Sixty-one—far from seventy, true, but not too far! I am a poet. I am a
writer. And no, I did not begin saving at age twenty and no, I do not
have enough savings and no, my social security is not adequate and
no, my mortgage is not paid off, and no, I will not be able to retire,
whatever that is, at age sixty-five. No way. This brings us to the subject
of yoga.

I always had the idea that when I turned seventy, I would start taking yoga. Life would afford me that. No time to go into decline. Time
to grow, to change, to stretch, to keep learning, to keep creating, to
walk into new worlds, to stand on one's head. Time enough and space
enough, for I would (somehow) be retired. Retired, not from creative
work, but from the stress and pressure of working many more than
forty hours a week editing and teaching and coaching to finance the
basic necessities, to live in a simple but decent manner. After a lifetime
of hard work we are supposed to get a rest.

Aren't we?

In 2003 I am a self-employed person who lives by herself in a house she owns in an expensive metropolitan area (Seattle). In U.S. Census terms I am one of 25 million Americans who live alone. I am one of 193,000 Americans employed as "Writers and Authors." I'm healthy. I'm happy. I'm a productive writer. I do the work I love, and I love the work I do. In 2003 I'm health-insured, unlike 60 million other Americans. And this is the rub. My health insurance costs twice as much as my shelter, and keeps me from saving much.

So here I am. My dear country with its most expensive and least effective health-care system in the industrialized world and my dear younger self with her financial ineptitude have collaborated to provide me with a working older age, with what could be a working old age. This being the case, as the Buddhists say, how shall I proceed? My mother's death was painful and sad and it was like crossing a turbulent river. Now I am on the other side.

I am making financial decisions, deciding directions for my creative work, redecorating my living room. I have decided to retire immediately. I will keep on earning my keep, of course. But I will do everything I ever wanted to do, right now. I will work more efficiently, increase the daily time of work on my own writing, see a film every week. I will take yoga.

I sign up for the Friday evening class at Seattle Yoga Arts, on 15th Street on Capitol Hill. I go to my first class.

The studio is a space of candlelight and dim incandescent light, silence, an expanse of polished maple-wood floor. Sticky yoga mats, blue and green, are stacked in one corner. The walls and baseboards are painted raw sienna and burnt sienna—earth colors of gourds and nuts, seedpods, pinecones. Two or three students sit in the lotus position, meditating. To the left of the door is an entry area with a worn carpet, benches, a coat rack, a water cooler with a half-full cistern, a small bookcase stuffed with books and papers. At the center of the far wall, in an alcove, candles flicker among statues—are they Krishnas? Shivas? Kuan Yins? Bodhisattvas? This is the altar wall. The ceiling is softened with lengths of pale orange linen. Harnessed into blue straps hooked to the ceiling, a man hangs upside down.

Our teacher, Lisa Holtby, greets me, asks if I have any injuries. She wears a black leotard and tights and a thick cardigan sweater. She has straw-colored bangs cut straight across and ruddy, high-boned cheeks. She looks more like an advertisement for Kansas or for Dutch Boy Paints than a yoga teacher. I tell her about my knee. Later, she stands before the class in bare feet and black leotard. Her palms press together in front of "the heart." She leads the class in a blessing (you can participate if you feel comfortable) and in the Om (if you feel comfortable). Then we begin the series of asanas (yoga postures) known as the sun salute. Lisa's movements are slow, solid, smooth, muscular, square rather than light or quick. Her voice soothes and nudges the class into pretzel twists, bends, balances. In the middle of the class someone's cellphone breaks into an aria. "How surreal!" Lisa says. "It's not my fillings, I swear!"

The weeks go by. Between doing my writing, and teaching my writing classes, I faithfully attend my yoga class. I am terrible at standing on one foot. Twenty yogis stand in the tree pose (one foot on the ground, the other curled up into the groin, arms stretched to the ceiling). I fall first to one side, then to the other side. (Does this mean, I wonder, that my life is out of balance?) Lisa kindly gives me hints but she calls no special attention to my plight. My falls make loud thumps as if one person in the room is practicing leaps instead of standing like a lone pine tree. I lose my footing, lose my footing again . . .

Outside of class I teach writing, and in class I watch Lisa teach yoga. She is a master teacher. To teach is to create a space, to create and hold open a container for learning, growth, and development. It is a space created by kindly encouragement, prods, attention to the class as a whole, attention to each individual in the class. It is a space created by personal authority, knowledge, focus, clarity. To teach well is to speak clearly. To see what is needed and to address that. I like the word "guide" as well as "teacher." Good teachers, good guides, try from time to time to put themselves into the hands of other good teachers, partly to rest and grow in a context provided by someone else, partly to observe skilled teaching.

Like writing, yoga is a process you enter in to. You commit to the process. Just talking about it has no effect. Just wanting to be a writer is like just wanting to practice yoga. Nobody says, "I only wish I could practice yoga." People who want to practice yoga practice yoga. But how often do we hear, "I want to write." "I have this idea that I could write." "I dream of being a writer."

I dream of standing on one foot.

The practice of yoga has a lot in common with the practice of writing. Especially does the practice of yoga have a lot in common with the practice of writing poetry. Both yoga and poetry have emerged from extremely ancient traditions. Both have numerous branches and permutations. Both are breath based. Both require the practitioner to bring attention to the present, to the matter at hand. Both benefit from a calm, self-compassionate, benevolent attitude. Both can be highly challenging. Both call for years of practice. I am a long way from knowing whether anything about yoga is as thrilling as realizing a new poem.

But I like yoga. I like walking into a world unknown to me, a world that has existed forever alongside my world like a fourth dimension. I like the muted colors, and the quiet of the studio. I like the spirituality of it—a tolerant spirituality. I like the way Lisa guides us at the start of the practice to think about clarity or to think about compassion for ourselves. I like the calmness of the yoga studio.

For to walk into the yoga studio is to walk into a cave of calm concentration. The silence is a calm silence. The yogis are taking off their street shoes, getting their sticky mats and laying them out. They sit on their mats, breathing to calm themselves, or stretching, bending, twisting. The movements are fluid, the voices low. The sense of calm is vast, like a physical thing, like ocean waves slow-lapping the shore or the rustle of leaves in a grove of birch trees. Candlelight flickers at the altar wall. Dim lighting softens the edges of the room, mutes angles, quiets the polished-wood floor. Calmness is dusky light, hushed voices, those warm earth colors. The lack of clutter is calming. Lisa is calming in her very presence, her leotard, her tights, her bare feet, her plain fingers and plain cardigan and unpretentious smile.

During the ten years before her death my mother could move less and less. Strokes and diabetes felled her. She could not walk. She could not feed herself, or turn over by herself. She, the genius of the family, the brilliant psychologist, could no longer speak well. She wore diapers and she curled in a wheelchair. I am looking at a photograph of her on the Heron Point dock on the Chester River, the river that flowed past my childhood. Here is my mother in her wheelchair surrounded by her smiling family. She is not smiling. Her face is an image of pure suffering.

In her last weeks, when she could barely swallow, when she could barely speak, when none of the nurses could understand her whispered words, when we stayed at her bedside at all hours, when we tried to calm her terrible anxiety, when we moved her every ten minutes, when we fed her ice cream and called the nurses whenever we needed to, my mother's words were simple, monosyllabic, continuous: "Help me. Help me."

I'm terrified of strokes. I'm terrified of not being able to turn over.

I'm terrified of legs bent stiff, toes curled into soles, fingernails cutting palms. I wake at night. I move my arms. My legs. My fingers. I spread my toes. I turn over just to make sure I can. I turn over again. In my yoga class I twist and bend. I try to stand on my head. I try to stand on one foot. Wherever my mother is, I'm sure she's not with me in my yoga class. My mother would not be caught dead in a yoga class. But I am here. Breathing. Falling, and falling again. My mother is gone. But she's here in my genes. She's here in my thoughts. She's probably past all surprise when it comes to me. She probably wouldn't be surprised even to see me twisting like a pretzel or standing on my head or hanging upside down or balancing on one leg, my arms branching up as if to embrace the universe.

23

Thirteen Ways of Looking at a Fur-Covered Teacup

> I believe that, in any society, the poet should be the exponent
> of the imagination in that society.
>
> WALLACE STEVENS
>
> It is the artists that do society's dreaming.
>
> MERET OPPENHEIM

1.

Wallace Stevens, American poet. Born October 2, 1879, in Reading, Pennsylvania. Composed the quintessential Modernist poem, "Thirteen Ways of Looking at a Blackbird," published 1917. Meret Oppenheim, Swiss artist. Born October 6, 1913, in Berlin. Created the quintessential Surrealist object, *Breakfast in Fur*, exhibited 1936. They were contemporaries, but they never met. So, what accounts for the strange synchronicities of their lives? They created culture and they carried culture. They both believed in the imagination, in art as redemption. And so do I. And since my neural networks recreate their images, their images, recreated, constitute part of my brain. What they were constitutes part of what I am. And what they were constitutes part of what we are.

2. Their Century

The year is 1900. Wallace Stevens is turning 21. The century ushers in Freud's *Interpretation of Dreams* (1900). Einstein's Theory of Relativity (1905, 1917). Darwin's long shadow. God's demise. Cubism (1907–14).

The Armory Show in New York: Duchamp's urinal (1913). Art's demise. Oppenheim's birth (1913). The Armenian genocide (1915). Chinese brush painting working its way into a Wallace Stevens poem:

Among twenty snowy mountains,
The only moving thing
Was the eye of the black bird.

The century ushers in World War I. Meret Oppenheim's father, a physician, attending Jung's weekly seminars. Jung, his notion of animus and anima, the male and female within each person. Oppenheim recording her dreams daily, from age fourteen. The Great Depression (1930s). Surrealism. Stevens, asked whether he intended his verse to be of use: "Perhaps I don't like the word useful." Oppenheim, rendering her teacup useless by covering it in gazelle fur. Oppenheim as a poet:

Quick, quick, the most beautiful vowel is voiding.

The century ushers in World War II. The Nazi genocide. The atom bomb. Hiroshima. Heisenberg's Uncertainty Principle: observers altering the observed. Subjectivity. Subjectivities. Stevens's belief that for nonbelievers brought up religious, poetry is the religion. Stevens, delivering a lecture in 1951 at the Museum of Modern Art (MoMA) titled "The Relations between Poetry and Painting." Stevens receiving the National Book Award. At midcentury, in 1955, the poet Wallace Stevens dies of cancer. Oppenheim, now forty-one, is entering her productive decades. Artwork as dreamwork, figure as archetype, bird as flight of fancy. Nineteen sixty-eight: Vietnam. Revolution in the streets of Paris. Second-wave feminism in the 1970s and its elevation of Oppenheim to icon. Dream as sphinx: at age thirty-six Oppenheim dreamt a half-full hourglass. Exactly thirty-six years later, in 1985, the artist Meret Oppenheim dies of a heart attack.

3. His World

I am thinking Reading, Pennsylvania, the coal-fired world of slate and brick that Wallace Stevens was born into in 1879. I am thinking the Schuylkill River, the coal-carrying Schuylkill Canal, horses clopping down stone-cobbled streets, coal trains, the Philadelphia & Reading Railroad, railroad bridges and iron tracks and the Reading Iron Co. I am thinking apothecaries, Dry Goods, saddlers, shoemakers, shirtmakers, harnessmakers, printers, house painters, blacksmiths, stonemasons, brickmakers, bricklayers, tinmen, tanners, tailors, farriers, cabinetmakers, carpenters, and cutlers. I am not thinking *poet*. I am thinking of Stevens's mother, Pennsylvania Dutch, reading the Bible every night to her five children. I am thinking of a saloon on every corner and a church on every *other* corner. An industrious, virtuous, religious, slightly inebriated town, a manufacturing town, urban center to coalfields and cow fields and steel mills, a town where commodities, not poems, are produced, transported, purveyed. I am thinking of Stevens's father, a lawyer descended from Pennsylvania Dutch farmers, writing to his twenty-year-old son, "I am convinced from the Poetry (?) you write your mother that the afflatus is not serious—and does not interfere with some real hard work." No wonder then, that Stevens would later defend the maleness of writing poetry in an essay titled "The Figure of the Youth as Virile Poet." No wonder he would assert, "The centuries have a way of being male." From whence he came, real men did real work and manufactured real things, useful things—not poems—and in the process, made real money. And so, no wonder he wrote, in defense of the imagination, in defense of the artist, "The Man with the Blue Guitar": "The man bent over his guitar, / A shearsman of sorts. The day was green." Shearsman: one who shears sheep, a farmer, a real worker doing real work.

4. Her World

Meret Oppenheim was the granddaughter of Lisa Wenger, a suffragette, a well-known writer and illustrator of children's books. Meret's aunt, Ruth Wenger, was briefly married to Hermann Hesse. The maternal, Wenger side of the family was Swiss. Meret's father was a German physician. He informed Meret that "Women have never done anything in art." At age nineteen, in 1932, she went to Paris to study art. At age twenty she had a passionate affair with Max Ernst. This lasted for a year, but thereafter she continued consorting with the Surrealists and other

Paris artists—Picasso, Dora Maar, et al. Her own output included drawings, collages, assemblages, and plaster models of sculptures. Meret was brought up Protestant, but in 1936 her family, due to its Jewish name, moved out of Germany to Switzerland to live with her grandmother and on her grandmother's income, since in Switzerland her physician father was barred from practicing. In Paris, Meret, forced to become self-supporting, brought in income by designing clothing and jewelry. She entered into an intimate relationship with Man Ray (who famously photographed her). She became infatuated with her fellow Swiss artist Alberto Giacometti, twelve years her senior. (Giacometti did not return her infatuation.) Her first solo show took place in 1936. Her work at this time, writes art critic Bice Curiger, exhibited "astonishing artistic maturity, not in the sense of 'consolidation,' but of extreme self-possession."

5. Making Art while the World Burns

Consider the year 1936. We are deep into the Great Depression, into the great diaspora of the down-and-out, mendicants spreading out across America, across Europe. In Europe, fascism looms. Oppenheim is twenty-three, rather beautiful, living in Paris, studying art, making art, vulnerable, subject to melancholy, determined to live "a life unshackled by social conventions." Once, at a Paris café, in a wacky Surrealist gesture, she peed into the hat of a haughty gentleman. The artworks she made that year included a fur-lined bracelet. After talking fur at a café with her friends Pablo Picasso and Dora Maar, Oppenheim fabricated a fur-lined tea service. André Breton invited her to participate in an exhibition of Surrealist objects, and *Breakfast in Fur* was exhibited in Paris, Alfred Barr purchased it for New York's Museum of Modern Art, and it was exhibited in MoMA's *Fantastic Art, Dada, Surrealism* (1936–37), where Wallace Stevens, very much involved with the visual arts, with the New York art world, undoubtedly saw it. On her part, Meret Oppenheim became an instant celebrity. Meanwhile Stevens, age fifty-eight, a three-piece-suited vice-president of the Hartford Accident and Indemnity Company, was writing poems on the train commuting from Hartford to New York, writing poems on his long solitary walks. In February 1936, on vacation in Key West, he got into a drunken brawl with Hemingway and whacked Hemingway in the jaw so hard that he, Stevens, broke his hand in two places. Stevens was a huge man, called "the giant," a successful insurance executive who succumbed to episodes of drunkenness. He spent numerous days and weeks away from Elsie, the Reading girl he'd married, who resented his literary career, who believed his poems were her poems, written for her (as at first they were), who felt their publication as betrayal. That year, 1936, Knopf published Stevens's second book of poems, *Ideas of Order*, and critics began to acknowledge him as a major American poet.

6. After the Teacup

In the late 1930s, the war coming on, Oppenheim returned to Basel, Switzerland, and lived at her parents' house, a small dwelling behind her grandmother's house, which was now being rented out. Oppenheim brought in income by designing clothes, and she learned the craft of art conservator. She had fallen into a crisis, a severe depression and loss of direction that would plague her for seventeen years, until 1954. She made art but in a directionless manner, stacking works in corners, destroying works. It was a crisis of "shattered confidence," as she later described it. She felt "as if millennia of discrimination against women were resting on my shoulders, as if embodied in my feelings of inferiority." Meanwhile, Stevens was hitting his stride, both as a poet and in the insurance business, while around him the world was falling to pieces. Stevens quietly and freely gave away money, supporting literary magazines and various artistic endeavors and persons. From a mutual acquaintance he learned that, in Europe, Hermann Hesse was in financial straits. He quietly arranged to buy some of Hesse's watercolors. He recorded in his notebook, "The world without us would be desolate except for the world within us."

7. Comments on the Teacup

In 1948 Stevens recorded in his notebook: "Some objects are less susceptible to metaphor than others. The whole world is less susceptible to metaphor than a teacup." In 1970 Oppenheim made a collage on paper, a kitschy image of a saucer, a stubby fur-colored spoon, and a fur-colored teacup. This she titled *Souvenir of "Breakfast in Fur."*

8. Meret Oppenheim Becomes Meret Oppenheim

In 1945 Oppenheim met businessman Wolfgang LaRoche, and in 1949 they married, remaining a couple until his death in 1967. In 1954 she regained confidence, established a studio, and during the next thirty years went on to make more than a thousand artworks. In 1975 Oppenheim stated: "I think it is the duty of a woman to lead a life that expresses her disbelief in the validity of the taboos that have been imposed upon her kind for thousands of years. Nobody will give you freedom, you have to take it."

9. Imago: Meret Oppenheim

Oppenheim, the art conservator, manipulated an eclectic range of materials to make a body of work full of dream figures, insects, personas, birds, clouds, masks, and snakes (to Oppenheim snakes represented "creative force, an attribute of female divinity, evolution, nature").

She made *Mask with Tongue Sticking Out* of wire mesh with overturned plastic bowls for eyes and nose, with the word "Bah!" inscribed on the pink-velvet tongue.

She sculpted the enigmatic *White Cotton Wool Mask* out of cotton wool with magical many-lashed wire-and-sequin eyes and a curling tail-length tongue made of wire and fabric.

She made a stiff, ominous-looking persona, *Octavia*, out of wood, plastic, and a tree saw.

She cut dragonfly wings out of tin, put a screw for the thorax, and mounted the insect *Dragonfly Campoformio* in a wood box.

She made *The Raven* out of oil paint and wood and molded fungus.

She made a terrifying swollen red head, *Oaf*, out of traditional gouache.

She made *Enchantment*, an otter swimming in red ochre suns and pale moons, an above crescent moon mirrored in a below crescent moon, from oil paint on cardboard glued on wood. The otter, made of oil paint on wood, is screwed out from the picture plane so that it swims freely in high relief among the suns and moons. *Enchantment* is not about enchantment, it *is* enchantment.

She made *At a Grave* with colored pencil. This tender, sad drawing shows trees, stones, two abstract figures on a leaf-brown ground. She made it and she named it and the next day she learned that Giacometti had died.

She made the erotic, vase-like, full-wombed, two-breasted *Primeval Venus* out of terra cotta and oil paint, with a little stalk of straw for a head.

She made *Old Snake Nature* out of anthracite coal, wire mesh, and rugosit. Rugosit may be related to Rugosa, a Paleolithic coral. The coils of a thick glittering black snake curl up out of a burlap-looking bag. The head, turned toward you, is white, with a strange blue-green eye, staring.

She painted a car muffler to make *Queen Termite*.

She used her own body, once an x-ray of her head, a memento mori showing skull and earrings and ringed finger bones (*X-Ray of M. O.'s Skull*) and once a photograph over which she sprayed a tattoo, using stencils (*Portrait with Tattoos*).

Portrait with Tattoos is a powerful imago-like image, a head-and-shoulders self-portrait with the face painted as if with war paint so that the figure appears like a warrior or a spirit from a dream. Imago: an image or condensed perception of a person, often parent, that we carry in our unconscious from childhood. This imago then, Meret's imago, is like a primitive image carried to us from the beginning of time, the imago of a witch, perhaps. It is not the imago of a supplicating muse, not the imago of a woman preparing supper, not the imago of a lover or a mother. No. This imago is fearful, singular, self-contained, almost dreadful, like the sudden vision of a god. It is the imago of a ruler, of a queen, regal in her bearing. It is the imago of an artist at the height of her powers, gazing at us with defiance. It is the primeval imago of a powerful creator.

10. Libra: The Scales of Justice

Stevens was born on October 2. Oppenheim was born on October 6. Libra, sign of balance, the scales of justice. A Libra "would make a good lawyer, judge, or politician." Or a good artist. The balance between receptivity and bull-headedness, between imagination and the critical powers, between dream and waking reality, between masculinity and femininity. Oppenheim believed that the "androgyny of the spirit and intellect" was the core dichotomy possessed by the artist, by herself. The balance between perception and reality. Between the inside world and the outside world. Stevens wrote a poem titled "Not Ideas about the Thing but the Thing Itself." Oppenheim strove to create works that were "not simply a picture of an idea, but the thing itself." Collapsing the dichotomy between image and object, between being and knowing, between making and being. An Oppenheim drawing, *Scales. A Head in One Tray*, shows a beam balance. The tray on one side holds a human head. The tray on the other side holds nothing, emptiness. The two sides are in balance, equal.

11. Wallace Stevens: Titles of Works

- Thirteen Ways of Looking at a Blackbird
- Gray Stones and Gray Pigeons
- The Man with the Blue Guitar
- Study of Two Pears
- The Blue Buildings in the Summer Air
- Yellow Afternoon
- Man Made Out of Words
- Mountains Covered with Cats
- Large Red Man Reading
- A Dish of Peaches in Russia
- The Poem That Took the Place of a Mountain
- Dutch Graves in Bucks County
- How to Live, What to Do

12. Meret Oppenheim: Titles of Works

- A Blackbird
- Snake and Black Stones
- Old Snake Nature
- One Person Watching Another Dying
- Three Black Pears
- Oh, Too Bad, I Eat Sorrow Out of Tin Cans
- Why-Why
- Well, We'll Live Later, Then
- The Ancestor with Two Noses, Called Bird-Egg
- Red Courage Hunts in the Woods
- Hat for Three Persons
- The Night, Its Volume, and What Endangers It
- Who Risks It, Who Tries Again!

13. Holy Magic

Stevens, the poet who loved painting, who searched for meaning within his own interior world. Oppenheim, the artist who loved words, who searched for meaning within her house of dreams. They stood for the imagination, for enchantment, for the man with the blue guitar. They created in defiance of the worlds they were born into. They created despite wars, despite practicalities, despite obstacles and depressions and difficulties. Stevens wrote of how poets piece the world together with holy magic. Oppenheim, asked how one of her paintings came about, replied, "Strange things happen on the moon."

24

Going to Portland

The idea of a journey—to one's own self, to the depths of
one's longing—is crucial here.

MICHAEL TUCKER

To be American is to hit the road. It's to head west or south, to take
off, to blow through town, to ride the rails to kingdom come. What
could be more American than the long so-long of the train whistle's
wail. Or so I think as Amtrak's Coast Starlight train pulls out of Seattle's
King Street Station—that decrepit brick depot—heading south to
Portland, Oregon.[1] I'm on the train, leaving town. It's early February, a
cold misty morning. The wheels thrum, the whistle blows. The coach
clickety-clicks past back lots, Burlington Northern train yards, Port of
Seattle cargo yards stacked with intermodal shipping containers. We
pass mud ditches, oily pools, blackberry-choked road banks. We pass
Terminal 46 on the deep harbor of Elliott Bay, the port's gantry cranes
perched on the quay like great orange birds preparing to lift off into
thick fog.

Out the dusty train window, I photograph America: Doug fir and
western hemlock rising up along the freeway, cars moving slower
than the train. A stalled freight train with its boxcars, flatcars, gondola
cars, hopper cars. Rusted auto bodies, dry grass, exit ramps, the ware-
houses and loading docks of Kent, Washington. We cross the narrow,

1. Between 2008 and 2013 the City of Seattle restored King Street Station to its
former grandeur, including ornate plaster ceilings and a chandelier to illuminate
polished terrazzo floors. It is decrepit no more.

slow-flowing Green River on a steel-truss railroad bridge. We pass the Boeing plant, grounded aircraft half-hidden in mist.

I'm riding the rails south into new territory—new to me—and I feel unburdened, light, as if I've just jettisoned that bulky, over-stuffed suitcase—my life. I've been working hard and I need this weekend of reading and reverie and going to the Portland Art Museum. I've always loved visiting a new city. New landscapes, new cityscapes seem to invite new possibilities. I gaze out at ditches and parked trucks and parking lots and think, Well of course this is nothing new. I might be the millionth traveler over this route. I've joined a procession that winds back and back and back.

The train passes over well-worn tracks connecting Seattle and Tacoma, rival towns, once railroad towns, sawmill towns. The tracks were once Northern Pacific tracks controlled by Tacoma interests, denounced by Seattle as the Orphan Road. If you wanted to go south from Seattle to Tacoma, you had to wait in a deplorable depot for a train with no particular arrival time and take your belated bumpy ride backward since in Seattle the Tacoma train had no place to turn around. This was in the 1880s. In 1889 Seattle burned to the ground in the Great Fire, upon which Tacomans chanted "Seattle, Seattle, Death Rattle, Death Rattle!" But Seattle, indignant and fanatic, surged ahead.

The train's rocking and clacking sets off my own train of thought. Time runs along like a train, smooth and irrevocable, at least *present* time does. But, within a given landscape, *past* times stand still. Past eras stack up like beads on a string or like lumber or shipping containers. The landscape the Coast Starlight is traversing, the wetlands and forestlands of Puget Sound, has supported generations going back at least twelve thousand years. The land, like an old face, bears the marks and scars of past lives. The route itself is a palimpsest—newer roads overlying older roads. The ur-road is a trail long-traveled by the Nisqually, Steilacoom, Puyallup, Chehalis, and Chinook, and later, by Hudson's Bay Company fur traders. Later still, two friends, African American George Bush (that's right) and Irish American Michael Simmons, hacked a road through the forest. In 1860 came the military road. In the 1870s Chinese laborers graded and tracked the route, which today parallels Interstate

5. Now we ride the tracks, mostly ticketed tourists on this passenger train that runs per a deal between Amtrak and wsDOT. People are eating, gossiping, reading, gazing out at passing scenery. A century from now we will all be gone. Some other train carrying some other escapee from some other life will be—for all I know—heading south to Portland.

I walk the swaying cars to the café. I buy a hotdog, French fries, a coke—mortal sins where I come from—and take them back to my seat. I relish delicious salt and sizzling fat.

We reach Tacoma, take on passengers, continue south, and we "let that lonesome whistle blow." We pass the blue haze of Commencement Bay. We pass restaurants newly opened on the old Tacoma waterfront, new docks interspersed with remnants of rotting piers. Another palimpsest—working waterfront overlying historic waterfront. The Port of Tacoma imports some two hundred thousand autos—Mazda, Mitsubishi, Suzuki—per year. It exports Weyerhaeuser wood chips sent to Japan's paper mills and yellow corn sent to Asian feedlots. It imports gypsum from Mexico, and shoes, toys, salt, seafood, tallow, frozen meat. The Port of Tacoma hums with container vessels, car carriers, tankers, barges, tugboats, trucks, and unit trains.

We tick along and suddenly, high above, the Tacoma Narrows Bridge appears, a sliver of silver suspended in mist. Below, on the floor of the deep, turbulent Tacoma Narrows, lie the bones of the old Tacoma Narrows Bridge, opened to jubilation July 1, 1940, gleefully dubbed Galloping Gertie for its joy-ride undulations, collapsed November 11, 1940. A brief, delirious, disastrous love affair. A decade later the new Tacoma Narrows Bridge opened. It incorporates, like a second marriage, everything learned from the storms of the first catastrophe.

The book I picked to read on my trip to Portland is Ivan Doig's *Winter Brothers*. Doig spent a winter on Cape Flattery on the Pacific coast of Washington to write a diary while reflecting on pages of James Swan's diary written on the Pacific coast of Washington a century before. Swan began as a Massachusetts ship fitter, but in 1850 jettisoned his East Coast life to go west. He never lived with his wife or children again. He ended up in Port Townsend, and then on Cape Flattery, living with the Makahs and recording the Makahs' seagoing, whale-hunting, red-cedar-bark way of life. Swan's diary is a staggering record—some two-and-a-half million words written over four decades—of the far Northwest as it was five lifetimes ago. Ivan Doig wrote *Winter Brothers* in the early 1980s, now twenty years ago. His book reflecting on Swan's book has itself sunk through layers of time.

You could simply do as James Swan did—leave one life and begin another. When I boarded the Coast Starlight, I threw off my Seattle life as if it were a moth-eaten old coat. I could just leave that coat behind, just not return. Not that I don't love my life, but that I have worn that coat for a long time. I could begin a new life in Portland, Oregon, where I have never been. What an American thing to do. To hop a train and ride the rails to a new life. The thought pleases me.

Now the Coast Starlight is tracing the route of the old Northern Pacific Railroad from Tacoma to Portland, retracing the route of another journey, taken on November 3, 1885, by two hundred or so Chinese people of Tacoma, expelled by a white mob, picked up by a Northern Pacific train and carried weeping and shivering south to Portland while behind them their dwellings burned. A journey undertaken may be wished for or it may be forced. I feel sobered and sad.

My wished-for journey continues south toward Olympia through country I've traced with my finger on a map and now learn with my eyes. We pass through the Nisqually delta, a low-lying grassy wetland at the mouth of the Nisqually River, past the Nisqually Rez, and then past Olympia—the train purveys no view of the domed Washington State Capitol. We continue south through low-lying country that my geology guidebook tells me has been a treeless prairie since prehistoric times.

Centralia comes into view, a coal-mining town founded by black pioneer George Washington and his wife Mary. Modest cottages sit in picket-fenced yards along wide, flat, tree-lined streets. We pass Duffy's Antique Mall, a patch of scrubby wetland, then house trailers and rusted cars, then the town of Chehalis. We pass a ditch furnished with a sofa vomiting its stuffing into the mud. Disgusted domesticity disgorging its dis-contents.

Hummocks of brown grass. A flat prairie and far to the east, the foothills and snow-crusted peaks of the Cascade Range.

And now the wide brown Cowlitz River, tributary to the Columbia, flows south along the tracks. Thin red alder trees line the riverbank. I snap photos out my dirty train window. We stop at Longview, that planned city built to serve what would become the world's largest lumber mill. We start up again. The whistle wails. The Cowlitz River keeps on flowing our way—south. On the far shore now, the eerie hourglass of the Trojan Nuclear Power Plant comes into view and then passes out of view.

Here is the great Columbia River, twelve hundred miles long, a river of rushing currents and wild waterfalls now tamed by dams, now slow-moseying among sandbars, shifting between shifty riverbanks. As Woody Guthrie sang, "Roll on, Columbia, roll on." This week follows the week of February 1, 2003, the day the *Columbia* space shuttle broke up upon reentering Earth's atmosphere. How we all feel the loss of those seven astronauts. The *Columbia* was named for the *Columbia Rediviva*, the first Euro-American vessel to enter the waters of the Columbia River. I feel the layers of time, journeys overlying journeys.

The great river divides Vancouver, Washington, from Portland, Oregon. We arrive at our destination (destined station) over the massive steel-truss Burlington Northern Railroad Bridge 9.6. We detrain into the historic brick Union Pacific Railroad depot. I am here.

I walk to my clean, inexpensive room at The Mark Spencer Hotel. I spend the weekend walking along the Willamette River admiring old bridges, browsing the world's largest bookstore (Powell's City of Books), sipping Peet's coffee, wandering through Chinatown, photographing the Chinese gate, the long riverfront park, a Chinatown establishment with the eerie name "Tacoma Café." I ride huge silent streetcars powered by overhead wires, humming along tracks over brick and cobble streets.

At the Portland Art Museum I stand before a human-sized male figure carved in basalt. It is more than a thousand years old. It has ribs, a navel, a distinct penis, a headdress or hat, an inward-looking face, slit eyes closed. It was found beside the great river and is believed to have stood at the center of a village. I feel a kind of power emanating from this god or ancestor or spirit. I feel in awe. Was this how his own people felt, living in his presence a thousand years ago? They are gone; a thousand years from now, I will be gone. Will I, a poet, have contributed anything even remotely to compare with this spirit-figure carved in stone? His power, it seems to me, exists outside of time.

I like being here in Portland, away from the responsibilities of teaching and editing. I like the solitude, the freedom to let my thoughts drift and turn. I like being on the road, "laughing, free, and gone" as that sappy old American road song would have it. I like the airy possibilities of being a new person in a new place. But standing here, standing before this mystical figure carved in stone, I am brought to a full stop. I wonder, Where am I going? What do I bring to what has gone before?

NOTE

Each of these chapters has appeared in earlier renditions listed in the acknowledgments. The chief difference between what you have here and earlier versions is that the science, wherever it comes onstage, has been completely updated. The chapter most affected by these renovations is "Genome Tome." The original version of "Genome Tome" appeared in *The American Scholar* in 2005, and in 2006 it received a National Magazine Award. Fifteen years have evaporated since I began composing "Genome Tome"; ten years since it received its glorious pat on the back. Here you have the second edition of "Genome Tome," structurally the same, personally the same, but in terms of science replete with what we know now.

ACKNOWLEDGMENTS

"Archaeology of Childhood" first appeared in *The Journal* and received *The Journal*'s William Allen Nonfiction Prize.

"Autobiography: An A-Z" and "Object & Ritual" first appeared in *Fourth Genre*, in vol. 15, no. 1 (Spring 2013): 115–12, and vol. 10, no. 2 (Fall 2008): 55–58, respectively.

"Banjo: Six Tunes for Old Time's Sake" first appeared in *Fugue*. It was reprinted in *Tribute to Orpheus: Prose and Poetry about Music or Musicians*, edited by Gary McKinney.

"Déja Vu" first appeared in *First Intensity*, edited by Lee Chapman.

"Genome Tome," first appeared in *The American Scholar*, published by the Phi Beta Kappa Society, vol. 74, no. 3 (Summer 2005): 28–41. Copyright by Priscilla Long. By permission of the publishers. "Genome Tome" received a 2006 National Magazine Award and was reprinted in *The Best American Magazine Writing 2006*.

"Going to Portland" first appeared in *Raven Chronicles*.

"Goodbye, Goodbye" and "Disappearances" first appeared in *Ontario Review*.

"Hildegard" first appeared in *The Chattahoochee Review*.

"Inheritance" and "Writing as Farming" first appeared in *North Dakota Quarterly*.

"Interview with a Neandertal" first appeared in *The American Scholar*, published by the Phi Beta Kappa Society (Spring 2011): 47–50. Copyright by Priscilla Long. By Permission of the publishers.

"Me and Mondrian" first appeared in *The Chariton Review*, in vol. 37, no. 1 (Spring 2014).

"The Musician," "Dressing," and "Balancing Act" first appeared in *Under the Sun*. "Dressing" was listed as "notable" in *The Best American Essays 2008*. "Balancing Act" was listed as "notable" in *The Best American Essays 2011*.

"My Brain on My Mind" first appeared in *The American Scholar*, published by the Phi Beta Kappa Society, vol. 79, no. 1 (Winter 2010): 20–37. Copyright by Priscilla Long. By permission of the publishers. "My Brain on My Mind" was listed as "notable" in *The Best American Science and Nature Writing 2011*.

"Notes Composed in the Dark of Our Time" first appeared in *Terrain.org: A Journal of the Built + Natural Environment*.

"On Quietness" first appeared in *Tampa Review*.

"Stonework" first appeared in *Passages North*.

"Solitude" first appeared in *Gettysburg Review*, vol. 21, no. 4, and is reprinted here with the acknowledgment of the editors. "Solitude" was listed as "notable" in *The Best American Essays 2009*.

"Thirteen Ways of Looking at a Fur-Covered Teacup" first appeared in *WebConjunctions*.

"Throwing Stones" first appeared as "Too Late for Miss Roselli" in *Pass/Fail*, published by Kleidon Publishing Inc.

I am grateful to the editors of journals who have read and appreciated my work and brought some of it to light. Robert Wilson, along with his perspicacious and jolly team at *The American Scholar*, stands at the top of the list. I warmly thank Phoebe Bosche and Anna Balint at *Raven Chronicles*, Lee Chapman at *First Intensity*, Heidemarie Weidner at *Under the Sun*, Simmons Buntin at *Terrain.org*, and, with special gratitude, the late Robert W. Lewis at *North Dakota Quarterly*.

I thank the following scientists for scrutinizing sections of this work and helping to make the science accurate and current (any mistakes, though, are strictly my own): Chris Tachibana, Ben Lansdell, Friedemann Schrenk, and Peter Kessler.

It is my good fortune to be part of the dedicated and ever-convivial gang that produces HistoryLink.org, the online encyclopedia of Washington state history. I lift my glass to you and to the memory of our beloved co-founder, Walt Crowley.

My third-Sunday workshop and performance group (the Seattle Five Plus One), now in its twenty-fifth year, has sustained me in my creative work over these many years. Love and special thanks to Geri Gale, Jack Remick, Gordon Wood, Don Harmon, and M. Anne Sweet.

My sister Pamela O. Long and brother-in-law Bob Korn have been insightful first readers for much of this work. I can hardly do without them.

Ms. James Kessler has put her eagle eye on these pages to their everlasting benefit. Elizabeth Wales was an immense aid in helping me to weave the chapters together. I thank the anonymous readers for the University of Georgia Press for their perceptive comments, as well as the kind and uber-competent crew at the press: Crux series editor John Griswold; director Lisa Bayer; and John Joerschke, Elizabeth Crowley, Amanda Sharp, David Edward Desjardines, and copyeditor Sue Breckenridge.

I thank my dear friend the painter Jacqueline Barnett for the image of her painting *Hope*, which appears on the front cover. Finally, Dr. Jay Schlechter has been a steady and loving support all along. Thank you.